RECALL!

California's Political Earthquake

Larry N. Gerston and Terry Christensen

M.E.Sharpe
Armonk, New York
London, England

Cover photographs on paperback edition courtesy of www.TrainWeb.com
(Gray Davis) and Ryan Balbuena (Arnold Schwarzenegger).

Library of Congress Cataloging-in-Publication Data

Gerston, Larry N.
  Recall! : California's political earthquake / Larry Gerston and Terry Christensen.
    p. cm.
  Includes bibliographical references and index.
  ISBN 0-7656-1456-1 (alk. paper) — ISBN 0-7656-1457-X (pbk.: alk. paper)
    1. Governors—California—Election—Case studies. 2. Elections—California—Case
  studies. 3. Political campaigns—California—Case studies. 4. California—Politics and
  government—1951– 5. Davis, Gray, 1942– I. Christensen, Terry. II. Title.

  JK8793.G465 2004
  324.9794′054—dc22

                                                                    2004041663

Printed in the United States of America

The paper used in this publication meets the minimum requirements of
American National Standard for Information Sciences
Permanence of Paper for Printed Library Materials,
ANSI Z 39.48-1984.

BM (c) 10   9   8   7   6   5   4   3   2   1
BM (p) 10   9   8   7   6   5   4   3   2   1

To the courageous people who run for public office:
Win or lose, you are leaders.

Contents

Preface   ix

Photos follow page 78

Part I. The Politics of Recall

1. The California Republic: More Like a Nation Than a State   3

Land of Direct Democracy   4

"California Sets the Pace": Tom Brokaw, *NBC News*,
October 8, 2003   7

California's Economy: Ranking Sixth in the World If
the State Were a Nation   9

Diversity Compounded by Diversity   11

Fractured Political Parties   12

Out of History and Diversity: Opportunity—and Opportunists   14

2. Elements of a Perfect Storm   16

The Power Crisis   17

A Plunging Economy   20

The Budget Deficit   21

And Then the Rains Came . . .   24

Part II. The Shaping of the Battle

3. The Legislature: Gridlock Defined   29

The Problem of the Two-Thirds Vote   30

The Term-Limits Paradox   32

Safe Districts and Dangerous Legislative Conflicts   35

Adding Fuel to the Fire   36

4. Governor Gray Davis: Triumph, Decline, Denial   38

Who Is Gray Davis?   38

Governor Davis: "Implement My Vision"   41

Slow Reaction: A Governor in Denial   45

Davis Agonistes   48

5. Qualifying for the Ballot: Right-Wing Conspiracy or Convergence of Interests?   49

Qualifying for the Ballot   49

Instigators and Visionaries   50

The Stealth Campaign: The Internet and Talk Radio   55

Darrell Issa's Golden Egg   57

Davis Fights Back   58

Republicans Engaged   59

An Election Is Ordered   60

Part III. The Campaign

6. The 75-Day Sprint   65

The Recall Goes to Court   66

A Two-Part Affair   67

Gray Davis: Defending Slippery Turf   67

Arnold Schwarzenegger: Capitalizing on the Mystique of Celebrity   71

Cruz Bustamante: The Democratic Alternative   74

Tom McClintock: The Conservative Conscience   76

Peter Camejo: The Liberal Conscience   77

Bill Simon, Peter Ueberroth, and Arianna Huffington: The Dropouts   78

The Supporting Cast   80

And the Winner Is . . .   81

7. The Issues   82

Voter Concerns   83

Candidate Positions   83

The Economy and the Budget   85

Education: We're All for the Kids   88

Energy: Crisis and Conspiracy Theories   88

Immigration—and Immigrant Candidates   89

Social Issues: Gay Rights, Gun Control, Abortion,
   and Women's Rights 91
Other Issues: The Environment, Health Care, and Political Reform 93
Issues: The Bottom Line 94
8. Interest Groups: Taking Sides 96
The Right: Conservative Groups and the Recall 97
The Left: Liberal Groups and the Recall 98
Labor: Rallying the Troops 100
Business: A Winner at Last 103
Indians: California's Newest Players 106
Follow the Money 107
Personality Politics: Movie Stars and Political Stars 108
A Different Sort of Election 109
9. The Campaigns and the Media:
Whose Election Is It Anyway? 111
The Campaigns: Shaping the Messages 112
The Polls: What Did We Know and When Did We Know It? 116
The Debates 117
The News Media: Kill the Messenger 119
Alternative Media: Celebrity Politics 123
The Final Cliffhanger 126

Part IV. The Outcome

10. Davis Loses the Recall Battle 129
A "Normal" Political Campaign in an Abnormal Political Climate 130
A Mandate for Change 131
Key Factors 132
The Schwarzenegger Surprise 134
Voting Groups 135
A Smooth Operation After All 138
Anatomy of a Defeat 139
11. Recall and Political Stability in California:
Can They Coexist? 140
Putting the Pieces Back Together 140
Lessons Learned 141

The Permanence of Structural Issues 143

California Recall: "Perfect Storm" or Trendsetter? 149

The Uncertainty Ahead 152

12. **Epilogue** 154

Out with the Old—In with the New 154

Schwarzenegger the Activist 155

Early Victories 156

Early Defeats 156

Early Compromises 157

The Movie Star and the Bully Pulpit 158

More to Come—But What? 160

Appendix A: Recall Timeline 163

Appendix B: The 2003 Recall Election and California Politics
on the Internet 167

Notes 169

Index 181

About the Authors 187

Preface

Welcome to California. You'd better have your passport in hand, because this is a different place from just about anywhere else you might go. California is the land of innovation, fads, and extremes. Check out almost any value or concept in the other forty-nine states and you may find it testing the limits and forging new boundaries in California. What you see here today, you will find in the rest of the country tomorrow. The tax revolt, environmental movement, and even anti-spam legislation began in California. And it is in this volatile, changing, strained political environment where the recall—a way for voters to yank elected officials from power at virtually a moment's notice—has taken root and blossomed.

Recall! California's Political Earthquake documents the latest chapter in California's exotic history, explosive present, and unpredictable future. Carried out in a state that is known for make-believe, the events behind *Recall!* are the true story of a polity that almost crashed under the weight of its own problems, a governor unable to manage public expectations, a public that simultaneously demanded answers to expensive questions but wanted them at no financial cost, and a cast of adversaries primed to take advantage of unique political, economic, and social circumstances.

California is not the first state ever to suffer from deficits, power outages, and a decaying infrastructure. But California has a provision in its state constitution that allows the voters, for no particular reason, to remove from office any elected official, thus making it easy to exercise public will on a mere whim. That incredibly low threshold for change was tested in the recall effort against Governor Gray Davis on October 7, 2003. As a result, California's "direct democracy" laboratory exceeded the wildest dreams of the reformers who brought the concept to the state nearly one hundred years ago.

Recall! California's Political Earthquake is about more than the recall event, however. It casts this incredible moment in the context of the state's political system, institutions, public officials, economic calamity, and celebrity. Only by understanding all these elements and more can one appreciate the public upheaval that defined the not-so-Golden State.

Was the recall good for California? Did the political bloodletting soothe the voters' anxieties while putting the state's remaining elected officials on notice? Read *Recall! California's Political Earthquake* and decide for yourself.

We couldn't have produced this book within weeks of the October 2003 recall election without a lot of help. First and foremost, we must express appreciation to the journalists who covered the election for their hard work and insights in a challenging situation. They are widely cited throughout this book. More than a source, however, our conversations with them throughout the election helped us develop our own questions and analysis. Producers and colleagues at NBC11 in San Jose were particularly supportive, especially reporter Beth Willon, who followed the campaign on the front lines. Along with other journalists, academics, and elected officials, we also thank organizations like The Field Poll, the Public Policy Institute of California (PPIC), and the *Los Angeles Times* poll for helping us understand what the people of California were thinking and for making their survey results freely available.

Our students and community also played a part by inviting us to talk about the recall election while it was under way and to analyze it when it was over. Neither they nor we knew we were often rehearsing parts of this book. Peter J. Tessier, a City Year Americorps member and student at San Jose State University, assisted with research and reliably found what we needed when we needed it. Elisa Gerston helped us focus on the big picture by patiently dealing with technical issues in preparing the manuscript. Ray Allen enthusiastically supported this project from start to finish, despite the necessity of putting his life on hold to meet our deadline. Heartfelt thanks to Elisa and Ray.

Finally, we are grateful to the excellent team at M.E. Sharpe. Our editor, Niels Aaboe, had the foresight to take on this project, the faith in us to complete it on time, and the determination to pull us all together to meet a tight deadline. He greatly improved the book with his meticulous, knowledgeable, and insightful editing. We also wish to thank our copyeditor, Heidi Thaens, and Angela Piliouras, Managing Editor at M.E. Sharpe.

For all the help and support we had, we acknowledge full responsibility for the book itself and any errors in it. We've done our best to tell the story of California's political earthquake fairly and accurately and especially to put it in a historical and political context that makes it understandable. We loved doing this—and we love California and its constantly changing politics. So our final "thank you" must be to California and its people.

Larry N. Gerston
Terry Christensen

Part I

The Politics of Recall

1

The California Republic

More Like a Nation Than a State

Every so often the earth moves in California. Sometimes the movement is mild—just a tremor—but at other times the land and its people are shaken by massive earthquakes. Such events are transformative. Cities are destroyed and rebuilt in new shapes and forms. People's lives end or are altered. Some flee to more stable places; others breathe a sigh of relief and get on with their lives.

But tectonic plates aren't the only things that move modestly or massively in California. So does politics in the Golden State. And every so often politics in California is shaken up as if by a massive earthquake. The latest shocker came with the recall of Governor Gray Davis.

It's not the first time that the state has been rattled. Think back to the transformation of California from an agrarian Mexican territory to the boom of the Gold Rush and statehood—almost overnight. Less than twenty years later the railroad arrived and built not just a transportation system but also a political machine that seized control of the state. The Workingmen's Party and the Grange battled the railroad machine in the nineteenth century and even wrote a new state constitution, but it was the Progressives of the early twentieth century who next truly shook up California with reforms that included primary elections, nonpartisan local elections—and direct democracy. Then came the rise to majority status of the California Democratic Party in the 1930s, only to be foiled by the long-standing popularity of Republican Governor Earl Warren. In 1958, Democrats shook things up when they finally elected a governor and brought two-party politics to California for the first time in the century. In 1978, the state was rocked by a voter revolt over property taxes, and then again in 1986, when voters removed three liberal justices of the California Supreme Court from office. In 1990, violent tremors shook Sacramento when Californians enacted term limits for state legislators.

3

Democrats elected only three governors in the twentieth century, until 1998, when they elected a fourth, Gray Davis, and at the same time won every one of California's statewide offices for the first time in 120 years. They also solidified their control of the state legislature. At the dawn of the twenty-first century, Democrats appeared to be invincible. But political prognosticators are no better at predicting massive change than geologists are at predicting earthquakes. Although a series of political tremors gave warning, few noticed and few predicted the "big one" that hit in 2003.

Then, over the course of just a few days in July 2003, California's recall election burst onto the world scene. California media had paid some attention to the petition process leading up to qualification for the ballot, but when the recall election became certain, national and then international media began reporting on it, although mostly just the serious newspapers. At this point, it appeared to be an interesting exercise in democracy, an entertaining embarrassment for Democrats—or both.

More people and more media began to pay attention when it became known that movie star Arnold Schwarzenegger was considering running as a replacement candidate for Democratic Governor Gray Davis. Rumors and speculation were rife, but Schwarzenegger kept his intentions to himself until he appeared on Jay Leno's *Tonight Show* and dramatically announced that he would be a candidate—contrary to the carefully leaked rumor that he would not run. The next day CNN ran a news special around the world—and the international and entertainment media of the world descended on California. The "big one" was about to hit.

How could it happen? How could it happen in California? How could an experienced politician who had been elected to statewide office no fewer than five times be challenged by a bodybuilder and movie star for leadership of the most populous state in the union with an economy that, if California were an independent nation, would rank sixth in the world?

Perhaps such a thing could happen only in California, but given the state's history and nature, it shouldn't have come as a complete surprise. The state flag proclaims the "California Republic," and in some ways California is more like a nation than a state. Political earthquakes have shaken California before and will again, but will this one radiate tremors and even quakes to other states and even other nations? The nation and the world watched the California recall with amusement—but also trepidation.

Land of Direct Democracy

California has long been a source of ideas, policies, and institutions that pop up elsewhere, as trends radiate from the state's constantly changing politics. Direct democracy has been a major source of these trends. While not invented in California, direct democracy reached its apogee there. The concept of direct democracy, especially for amendments to state constitutions, goes back more than one hundred years. South Dakota was the first state to adopt the initiative and referendum—enabling voters to make policy or rescind acts of the legislature—in 1898. A dozen other states followed soon after. Oregon was the first to adopt the recall of elected officials in 1908. California adopted all three tools of direct democracy in one sweeping package in 1911.[1]

In some states, direct democracy was seen as an extension of democracy—a form of statewide town meeting. In California, however, the enactment of direct democracy was more a tactic in the long battle against the political machine than the flowering of pure democracy. Governor Hiram Johnson and the Progressives sought to make sure that even if the still-powerful Southern Pacific Railroad machine gained office again, citizens would have the tools to make policy, rescind government actions, or remove corrupt officials from power between elections. In the next few years, several other states followed California's lead.

No one is more responsible for the introduction of direct democracy to California than Hiram Johnson. The Progressive reformer gained his fame by taking on the much-feared and powerful Southern Pacific Railroad, which had used its economic prowess to dominate state politics. Rejecting what he perceived as an unholy alliance among big business, political parties, and elected officials, Johnson advocated tirelessly for a government where the people could have the final say. His zeal catapulted him to the governor's office where, in his inaugural address, Johnson stated that a successful government "must rest primarily on the rights of men and the absolute sovereignty of the people."[2] Passionately advocating that every citizen be "his own legislature," Johnson called on the legislature and the people to adopt the concepts of direct democracy—the initiative, referendum, and recall. The initiative enables citizens to make policy directly by securing signatures from the voters in order to place issues before them (for laws, 5 percent of those who voted in the previous gubernatorial election is required; for constitutional amendments, 8 percent).[3] The referendum allows voters to overturn acts of the legislature. Referenda qualify for the ballot when petitioners attain a number of signatures equivalent to 5 percent of participants in the most recent gubernatorial election. Finally, if sponsors collect enough signatures to place the issue on the ballot, the recall allows citizens to remove elected officials from their positions.

But while other states have also adopted direct democracy, none have used its mechanisms more frequently than California. Referenda and recalls have been rare, but initiatives have become commonplace in California. Between 1912 and 1977, 159 initiatives qualified for the ballot, although only 43 were approved. Then, in 1978, Californians rediscovered the initiative with Proposition 13, the property tax reform initiative that launched voter revolts around the country and around the world. Angry voters, organized interest groups, candidates on the make, and political consultants latched onto the initiative, placing 132 measures on the ballot between 1978 and 2003; of these, 55 passed.[4]

Among them were many measures restricting the taxing powers of state and local government in California—the progeny of Proposition 13. California became known for "ballot box budgeting," a phenomenon that tied the hands of governors and legislators facing fiscal crises. Noting the success of these efforts in California, other states followed suit. But beyond the use of the initiatives, California had set off a taxpayer revolt that spread to states and nations lacking the option of direct democracy. Antitax sentiment swept the country and the world.

Meanwhile, back at the ballot box, Californians busily set other national and international trends throughout the 1990s. California was among the first states to adopt term limits for state legislators in 1990, cutting their staffs and other resources at the same time; the term limits movement soon spread across the country, almost always by initiative. In 1994, California voters approved Proposition 187, an initiative restricting government services to illegal immigrants. Although the federal courts ruled much of the measure unconstitutional, Californians (not for the first time) expressed their anti-immigrant sentiments and put the immigration issue on the political agenda. That same year, Californians enacted the "three strikes" initiative, imposing strict and lengthy sentences on those who commit three felonies. Other states and the federal government soon adopted the three strikes rule. Two years later, Californians stoked the national debate about affirmative action by approving Proposition 209, an initiative to eliminate affirmative action by California state and local governments. A ban on bilingual education followed in 1998, again a forerunner to a national debate. California is still locked in combat with the federal government over the medical use of marijuana—approved by California voters in 1996 and in other states since then but still banned nationally.

As for the recall, until 2003, this device had been used almost exclusively against local government officials. Depending on the office, anywhere between 10 and 30 percent of the voters are required to sign petitions over periods of time ranging from 40 to 160 days. Historically, recalls have been used against school boards more than any other group of elected officials. The recall has been used successfully on state legislators only four times in

6

its entire history, and no member of the state executive branch had ever been recalled until 2003.

California governors have not been exempt from recall efforts, however. Thirty-one previous attempts have failed to collect the necessary signatures— 12 percent of those voting in the previous gubernatorial election. None even came close to qualifying for the ballot. Based on the 12 percent requirement, recall proponents in 2003 needed 897,158 valid signatures. Most experienced political observers doubted that such a number could be reached. But while the requirements for recalling a statewide official in California seem intimidating, it should be noted that the state's 12 percent threshold is the second lowest of the eighteen recall states, perhaps a source of encouragement to proponents in the early days of signature-gathering. Montana has the lowest requirement with 10 percent. Of the seventeen remaining recall states, all but three require 20 percent or more of the voters to sign petitions; most require 25 percent or more, and Kansas requires 40 percent. California also makes recall somewhat easier than some other states by requiring no specific grounds—such as corruption or malfeasance.[5] Californians can recall officials just because they don't like them. But whatever the standard, the recall of statewide officials is extremely rare in American politics. Only one governor, North Dakota's Lynn Frazier, had been recalled before 2003, and that was in 1921.

The initiative, referendum, and recall have always been powerful potential tools of direct voter involvement in the political process. But whereas the initiative has become a staple of California politics, the recall was more a threat than a reality. That tool was activated in 2003, changing forever the rules of political engagement in California politics.

"California Sets the Pace": Tom Brokaw, *NBC News*, October 8, 2003

But direct democracy, ballot box budgeting, and the taxpayer revolt aren't the only trends California helped launch.

In the 1950s and 1960s—an era of infrequent use of direct democracy— Californians built themselves a great state and a powerful economy with massive investments in infrastructure and education. California constructed water and transportation systems (assisted by federal funds) to provide the infrastructure of the state's great cities. California created the most extensive and affordable system of higher education in the world, from virtually free community colleges to the multiple campuses of the California State University and University of California. Other states and nations followed California's lead and some have since surpassed the Golden State.

7

Other trends, too, started in California in the 1960s. Race riots in Watts, the black ghetto of Los Angeles, were among the first in the nation, setting off an era of racial turmoil in America's urban centers. Student unrest over the Vietnam War at the University of California's Berkeley campus spread to other campuses and other states. California's Cesar Chavez introduced farm workers and their plight to the world. And in 1966, partly in reaction to these events, California elected Ronald Reagan governor, launching a career that took him to the White House and revived the Republican Party nationally after the debacle of Richard Nixon (another Californian). Reagan wasn't the first actor to succeed in politics (U.S. Senator George Murphy preceded him), but he showed the world how far an actor could go—and others followed, from Clint Eastwood to Sonny Bono.

California has set trends for the country and the world in many other areas as well. Abortion, for example, has been a battleground in California for decades, with pro-life religious conservatives pitted against pro-choice women and their allies. While abortion is still a battleground issue in Congress and nationally, pro-choice candidates—usually Democrats—have consistently won statewide elections in California. Successful Republicans, such as Pete Wilson and Arnold Schwarzenegger, have been moderates on abortion. As a consequence of these attitudes, California's constitution and civil rights statutes provide more protection of reproductive freedom than U.S. law and policy, including a specific constitutional right to privacy, public funding for abortions for low-income women, funding for family planning, and access to abortion for minors without parental approval. On related issues, the state passed legislation authorizing stem cell research and providing workers with paid family leave—again setting trends.

California has been an even more powerful trendsetter on the environment, banning offshore drilling (over federal objections), setting aside open space and parks, and most notably setting standards for air quality. The Golden State virtually invented smog, but Californians are acutely conscious of its dangers. And even as Californians commute and pollute, they are passionate environmentalists. Earth Day was hatched in California in 1969—before much of the world had even heard the word *environmentalism*. In recent years, California has set more stringent standards for automobile emissions than the federal government—over the vehement objections of both automakers and the federal government. But the state's market is so huge that automakers have chosen to comply, making California's standard, in effect, the national standard. In 2002, California enacted legislation requiring automakers to reduce greenhouse gas emissions from their cars and trucks—ahead of other states and the federal government.

California has also been a trendsetter on gay rights, electing gays to public office before of the rest of the country and enacting nondiscrimination legislation and domestic partners' rights. Vermont jumped ahead of California by creating civil unions, however, and Californians joined the opposing national trend to restrict marriage to a man and a woman by approving an initiative in 2000. Even so, legislation was approved in 2003 to further extend the rights of domestic partners.

The inventory of trends launched in California could go on, although in the 2000 and 2002 elections California clearly did not follow the nationwide trend to the right. As George W. Bush narrowly won the presidency and Republicans increased their majorities in Congress, California went in the opposite direction, voting for Al Gore for president and, in 2002, giving Democrats their most sweeping victory in California history. But if California was out of step in those elections, the 2003 recall seems to have brought the state back into trendsetting mode. Once again, California is setting the pace. But will others follow?

California's Economy: Ranking Sixth in the World If the State Were a Nation

If California were an independent nation, its economy would rank sixth in the world, with an annual gross national product of over $1.3 trillion. The scale of California's economy is matched by its diversity, a source of strength even in times of recession. As one or sometimes more components of the state's economy decline, others remain stable or even surge.

Agriculture is one of the state's bedrock industries, dominating California's economy for most of its early history. The state remains the nation's top food supplier and major exporter of farm products, but other industries have dwarfed agricultural employment and revenues for a century. Oil and entertainment, for example, emerged as major forces in the economy of Southern California early in the twentieth century. Manufacturing—from cannery machinery to cars, airplanes, and computers—soon joined these traditional industries. During World War II and the Cold War that followed, military manufacturing grew exponentially in California—ships, airplanes, missiles, tanks, and other armaments became mainstays of the state's economy. At the peak of Cold War military spending, California received more than 20 percent of federal defense funds and was home to one-third of the country's top defense contractors, employing half a million people.[6] High tech grew out of the defense industries, emerging in the public consciousness as Silicon Valley in the 1970s and 1980s and booming in the 1990s.

But even with all this manufacturing, California's economy shifted from industrial to postindustrial. By 2002, service industries, including tourism, employed more than twice as many people as manufacturing. Agriculture, forestry, fisheries, mining, construction, and manufacturing combined accounted for about one-fifth of California's employment and gross state product. The balance of employment and dollars generated was in services, wholesale and retail trade, finance, insurance, real estate, transportation, communications, utilities, and government—a distinctly postindustrial economy.[7]

This happened partly because manufacturing was hit hard by recessions in the early 1990s and again in the early 2000s. California's economy slumped during the national recession of the early '90s, but the end of the Cold War brought a harsher blow. Federal defense spending in California shrunk by half and over 175,000 defense-related jobs were eliminated.[8] Southern California, with a heavy concentration of defense and aerospace industries, was particularly hard hit, and the economy languished for a decade. But economic diversity saved California, as the entertainment, telecommunications, and high-tech industries surged through the 1990s.

Silicon Valley, located between San Jose and San Francisco, led an unprecedented nationwide economic boom. At their peak, high-tech industries employed a million workers. In 1996, Silicon Valley became the top exporting region of the United States. High-tech industries substantially increased state tax revenues through jobs, sales, and capital gains as thousands profited from stock ownership.

But the boom went bust in 2001 and California and the nation slipped into recession. The terrorist attacks of September 11, 2001, stalled international high-tech trade at about the same time the dot-com bubble burst. Hundreds of dot-com startups (Internet-related companies selling goods, marketing information, or data management) disappeared. Leading high-tech companies like Sun and Silicon Graphics, Inc.(SGI), suffered serious setbacks and reduced staffing. Stock values plummeted, unemployment skyrocketed, and state tax revenues dropped dramatically. By 2003, unemployment hit 6.5 percent statewide and 8.6 percent in Silicon Valley, where 200,000 jobs were lost.[9]

Globalization of the economy benefited California at first, greatly increasing trade and exports for many of the state's products. Demand for workers was so great that many immigrated to California illegally, while industry sought highly skilled workers through a special visa program. But when the boom went bust, many of these workers were stranded. Globalization also resulted in huge outsourcing of California industry to other countries. American companies kept their headquarters and most of their research and development work in California, but they soon began exporting other work to countries with highly educated workers who could be employed at much lower cost.

10

Ultimately, California paid a high price as businesses and jobs left the high-cost state for greater profits elsewhere—not only abroad but also in the United States. At the beginning of the new millennium, California found itself in a deep recession, hemorrhaging jobs and developing a reputation for being unfriendly to business. The boom had been better than anywhere; the bust was worse than anywhere, and Californians were shocked that a state so rich could fall so far so fast. At the height of the boom in 1999 and 2000, surveys reported that over 60 percent of Californians thought the state was "going in the right direction." By August 2003, some 66 percent thought it was going in "the wrong direction," and the state of the economy had become their top concern.[10]

Like the proverbial time bomb, Californians' outrage over their economic misery would explode in political terms—it was only a matter of time. With the recall tools at their disposal, California voters acted.

Diversity Compounded by Diversity

The economic diversity of California is often a source of political conflict as workers struggle with business over health insurance, farmers fight urban developers for water, old businesses pay lower taxes than new ones, government regulations drive businesses to other states, and cheap labor draws employers to other countries. But with a population of 35 million, economic diversity is only one layer of California's complex diversity.

Racial and ethnic diversity are even more in the consciousness of Californians living in a state that is more diverse in this sense than any other state and almost any other nation. A constant flow of immigrants created this diversity and continues to feed it. Today, 26 percent of Californians are foreign-born. The state has no racial majority, although non-Hispanic whites remain the largest group, with 46.7 percent of the population. Latinos are 32.4 percent, Asians 11.2 percent, and African Americans 6.4 percent.[11] Most Latinos are from Mexico, but many are from Central and South America. Diversity among Asian groups is so great as to make generalization impossible—Japanese, Chinese, Koreans, Vietnamese, and more recently many immigrants from the Indian subcontinent. All of these groups are gaining political influence, but their interests vary considerably—both within and between groups. African Americans, Latinos, and Southeast Asians are more likely to be poor, uninsured, and unemployed or underemployed. Japanese, Chinese, Koreans, and Asian Indians do better. With the exception of African Americans, these groups do not necessarily vote as a bloc—and they are constituencies to which candidates and both major political parties are paying increasing attention.

11

But California's diversity does not end with economics, race, and ethnicity. Consider California's regions. The San Francisco Bay Area has long been culturally, economically, and politically different from the rest of California, with San Francisco as the financial capital and nearby San Jose as "the capital of Silicon Valley." Southern California is also distinct, with a different history and economic base—and with Los Angeles County distinct even from Orange County, San Diego, and the rest of the Southland. The adjacent "Inland Empire," centered on rapidly growing Riverside and San Bernardino Counties, has emerged as a force in its own right in recent years. While the Bay Area and Southern California have historically dominated the state, the vast Central Valley, running from Sacramento through Fresno to Bakersfield, has grown to almost equal stature. And distinct from all these urbanizing regions are rural areas in the deserts of the south and the mountains of the east and north. Even within these great regions, diversity is compounded by urban versus suburban versus rural interests.

Lifestyle and cultural differences also compound California's diversity. Religion, for example, reflects racial and ethnic differences, but it goes beyond that. California has the image of a tolerant, anything-goes state. But it also has a reputation for bizarre religious cults, from Aimee Semple McPherson to Jim Jones and beyond. And while many Californians practice no religion, others are passionately involved in their churches, mosques, temples, and synagogues. Conservative Christians and others expressed themselves politically on social issues from abortion to gay rights, becoming so active in the California Republican Party that they dominated a majority of its county central committees at one point. At the other end of the lifestyle spectrum are yuppies and dinks (dual income, no kids) in lofts in San Francisco and L.A. or in condos on the beach, along with gays and lesbians in West Hollywood and the Castro in San Francisco. They're a political force, too.

Fractured Political Parties

All this diversity is expressed in politics—sometimes consistently, sometimes sporadically. Even in politics, California is arguably more diverse than most other states. Organized interest groups, for example, speak for all of the diverse interests discussed above and many, many more. California has two major political parties—with considerable diversity and disagreement within each as Democrats challenge Democrats and Republicans challenge Republicans. California also always has several smaller parties; they rarely win elections, but they help express the diverse views of Californians.

If California is setting a trend in the operations of political parties, however, the direction is away from the more disciplined parties that operate in some other states and many other countries. Such parties assert strict discipline in assuring that candidates and officeholders toe the "party line" on issues, following agreed-on party platforms or the direction of party leaders. Strong parties recruit and support candidates, and their support is essential for political survival wherever they exist.

That's not the case for California's loosely organized parties. California's state party organizations try to recruit candidates, but the voters determine party nominees through primary elections, so mavericks can run and win. The state parties develop platforms on issues, but candidates and party members readily take alternative stands. California's parties raise money for campaigns, but interest groups contribute far more, and rich individuals—from Michael Huffington to Darrel Issa and Arnold Schwarzenegger—need not depend on party funds. Party organizations also try to keep order, framing the positions of their candidates versus those of the others in election campaigns. In simple terms, they provide a sort of loose political netting for candidates and voters. But the netting in California's political party system is so frayed that candidates and voters know they can't count on it.

Like those of many other states, political party organizations in California have little real control over what candidates say or do. But California differs from most other states in at least two distinct ways. First, the state's two major political parties have strong ideological components; Democrats have a history of ultraliberalism and Republicans have a history of ultra-conservatism. Both parties have moderates, especially among the voters who support them. But party activists tend to the ideological extremes, perhaps because such beliefs are necessary to motivate their activism. These ideological orientations are most strongly reflected in the state legislature, where members of both parties tend to the extremes and moderates are not merely an endangered species but have become extinct. Without moderates, compromise is impossible; hence, the deadlock on the budget and other issues.

But these attitudes extend well beyond the legislative arena. Both Democrats and Republicans have suffered from internal tensions. During the 1960s, the Democratic majority lost control of the state because of bitter infighting over Vietnam and civil rights. More recently, the Republicans have failed to capture key elected offices because of divisions on abortion, gun control, and immigration. Moderates in both parties seek the center, but the more ideological activists tend to dominate, both at party conventions when policy is determined and in primary elections when party nominees are chosen.

Internal tensions affected both major parties in the recall election as well. Democratic leaders tried to prevent any major Democrat from running to replace Gray Davis in the hopes that the lack of a Democratic alternative could save the governor. They ended up hamstrung by the candidacy of Lieutenant Governor Cruz Bustamante, who chose the clumsy theme, "Vote no on Recall, Yes on Bustamante." Clearly, Davis would have had fewer problems without that kind of competition. As for the Republicans, the large number of candidates seeking to replace Davis caused tremendous friction within the party organization; had the four major Republican candidates (Arnold Schwarzenegger, Bill Simon, Peter Ueberroth, and Tom McClintock) coalesced behind a single candidate, the Republicans would have had a much more dynamic campaign with consistent core messages.

A second difference between California and other states is a product of the reformers more than a century ago: the California State Elections Code has an extremely low threshold for placing new political parties on the ballot. All a new party has to do is register 1 percent of the number of people who voted in the last state election and *voila*! its candidates can be on the next ballot. A political party stays on the next ballot if at least one of its candidates for statewide office captures as little as 2 percent of the vote or if it retains a minimum number of registered voters. In 2002, along with the Democrats and Republicans, the Green, American Independent, Natural Law, and Libertarian parties all secured enough votes to qualify for 2004. Their presence has served to siphon away voter support from candidates in the two major parties.

These two factors—divisions within the major parties and the proliferation of smaller parties—provided a colorful backdrop to the 2003 recall election. As Davis was attempting to fend off all comers, one of them was his own lieutenant governor. Likewise, Republicans provided enough top-tier candidates to field a basketball team. Add to these the minor party and independent candidates and the 135-name list became something to behold. No wonder the contest took on the look of a political circus—there was an act for everyone!

Out of History and Diversity: Opportunity—and Opportunists

While the Great Recall of 2003 shook California mightily, it wasn't the state's first great political earthquake and it won't be the last. From the Gold Rush to the Progressives and on to Proposition 13, California politics has been transformed by dramatic events. Such change is almost built into California politics through its structures of government—especially direct democracy—its political traditions, and its amazing diversity. As Dorothy said to her dog in *The Wizard of Oz*, "Toto, I have a feeling we're not in Kansas any more." While divisions exist in every state, none has such complex and multilayered diversity as California, which means that a group of people can be found to advocate almost any idea, no matter how sensible or crackpot. Direct democ-

14

racy gives them an opportunity that doesn't exist in many other states. For a sample of such ideas, check out the initiatives under circulation as reported on the web page of the California secretary of state (www.ss.ca.gov). California's extreme diversity is expressed there, with a wide range of groups and interests taking the opportunity to put something before the voters. A current proposal would regulate the size of storage crates for pregnant pigs.

Most of these measures do not qualify for the ballot, however, and most of those that do are rejected. That's what a lot of people thought would happen when a little group of activists started circulating petitions to recall California Governor Gray Davis in February 2003, just weeks after he was sworn in for his second term in office. Another little group with another crazy idea. But the little group grew and the idea took hold and the governor, ultimately, was recalled. What happened wasn't exactly predictable—except to the extent that things like this can happen in California because of its very nature.

Was California starting another trend with its recall? The attention of the national and international press suggested it might be, although reaction outside California was more often amazed amusement than fear of a plague of recalls. If people elsewhere understood California better, they might not have laughed so much. But will California have the last laugh? Will other places experience similar voter revolts? Or was California's case unique—the result of a particular and peculiar set of circumstances? To mix metaphors (with apologies), was the California recall election of 2003 an earthquake that will send tremors throughout the country and even beyond, or was it the perfect storm?

2

Elements of a Perfect Storm

Governors are the most important of all state officeholders. In more cases than not, they enjoy veto power over legislative decisions, mold the courts by their nominations of judges, and possess superior access to information through management of the bureaucracy. Simply put, governors are at the apex of state power because of the clout associated with their offices—an endowment matched nowhere in American politics other than the U.S. presidency. No wonder that in 2002 alone Frank Murkowski left the U.S. Senate to become Alaska's governor, while Rod Blagojevich and Robert Ehrlich, Jr., gave up seats in the U.S. House of Representatives to become governors of Illinois and Maryland, respectively.

With so much authority at their disposal, governors can initiate, promote, oversee, and control public policies. One well-known book about governors concludes that their ability to manage issues ranging from preschool education to care for the elderly "predicts their chances for advancement" to higher office, perhaps the presidency.[1] This is especially true in California where, with 55 of the nation's 538 electoral votes, the governor is automatically considered presidential timber.

Likewise, failure to manage a state's pressing problems can spell difficult times for a governor. Such a predicament overcame California Governor Gray Davis. Within a period of 2 short years, the state was ravaged by an electrical power crisis, a suddenly depressed economy, and a budget deficit larger than the entire budgets of sixteen states. Regardless of whether these disasters were predictable or avoidable, the responsibility for solving them fell to Governor Gray Davis—at least in the eyes of the public. Whether Davis handled them well became more than a debatable question—it became a theme of the recall election movement. In a state known for bizarre movies with unpredictable endings, it would be hard to imagine more perfect elements for a storm of political revolt.

The Power Crisis

When things go wrong, most people care more about fixing them than how or why the problems developed in the first place. That's how it was when California's lights went out in 2000, bringing businesses to a halt and causing near panic among the state's residents. But it would be a disservice to mention the power crisis of 2000 without saying a word or two on its origin.

In 1996, the legislature deregulated electricity in California. Until then, three major utility companies controlled the production and sale of electricity to most of the state.[2] Authored by then-Republican Assemblyman Jim Brulte and signed into law by Republican Governor Pete Wilson, the bill had broad bipartisan support to place the sale of power in hands other than those of the parties producing it. The policy was very much desired by utilities, as evidenced by the more than $7 million in campaign contributions to legislators voting for the bill. Within the legislature, sponsors believed that if power distributors could buy electricity from several sources, they would be able to negotiate lower prices as a result of competition. The new law also required anyone selling electricity to reduce consumer prices by 10 percent.

Toward the end of the decade, the California economy, which had languished through most of the '90s, began to recover in earnest. Job growth expanded by 3 percent annually and the demand for power during hours of peak use soared,[3] forcing the state to import more than 25 percent of its power. Adding to the mix was the absence of any new power plant construction in California between 1988 and 2000—this as the state's thirst for power exploded. In June 2000, large regions of California experienced a power blackout; additional outages came close to occurring several times over the next few months, and rolling blackouts were ordered in December to control the flow of the small amount of electricity that was available to the state.

Although local energy companies selling power were restricted by price ceilings, those who provided the power were not. Suddenly, out-of-state wholesalers were selling electricity to California companies at exorbitant rates, depending on the time of day, the state's needs, and the amount of available electricity for sale. Whereas California power companies paid an average of $31 per megawatt-hour for the electricity they purchased in 1999, they sometimes paid as much as $1,000 per megawatt-hour during the year 2000.[4] Thus, during 2000 alone, state energy companies paid nearly $12 billion more for electricity than they were allowed to charge. Suddenly, they threatened to go into bankruptcy, bringing chaos to the state's aging electrical grid and distribution system. In the end, Pacific Gas and Electricity, one of the "Big 3" (along with Southern

17

California Edison and San Diego Gas & Electric), did go into bankruptcy, sending chills down the power spine of the rest of the state.

As the power crisis unfolded, Governor Davis was anywhere but ahead of the curve. In September 2000, Davis promised to bring all of the electricity players together in hopes of fashioning a solution to the crisis. Autumn came and went without the meeting. Davis seemed almost to throw his hands in the air at year's end, when, blaming Republican authors of the deregulation bill, federal authorities, and out-of-state power producers, he said "I don't have the legal authority to deal with all of the actors."[5] Meanwhile, as the crisis was growing, Davis raised $21 million for his reelection, including more than $500,000 from energy utilities. This approach would haunt the governor in the months and years to come.

Although the governor was not part of the original combination of factors that created the problem, he was nevertheless expected to find a solution— that's what governors are supposed to do when confronted by crises. On December 16, 2000, Davis finally called a special session of the legislature to take place in early 2001. Even then, he demonstrated little vision. "The rule is simple on special sessions [of the legislature]," a former assembly leader said. "When the governor calls you in, he's supposed to deliver a plan. Instead, Davis said, 'What do you want to do?' We were flabbergasted."[6]

With available electricity now daily at precariously low levels, the legislature and Davis struggled to find an answer. The first part emerged in February 2001, when the legislature approved Davis's request to authorize the state to purchase $10 billion of electricity through long-term contracts. Within a few weeks, the legislature issued another $10 billion in bonds to finance future power plants and return to the treasury the $50 million per day that had been paid to out-of-state power sellers because of the inability of local utilities to make any more purchases. Davis also used his executive authority to cut the red tape that was slowing down the construction of new power plants, and he asked the state's residents to embark on a major energy conservation program. Consumers responded surprisingly well, but despite reduced demand, supplies were still inadequate and expensive.

The fact remained that someone had to pick up the tab for California's soaring energy costs. In January, the state Public Utilities Commission (PUC), the agency responsible for electrical rates, approved a 9 percent rate hike across the board. Three months later, Davis proposed increases averaging an additional 27 percent; these, too, were approved quickly by the PUC. Suddenly, residential electricity users found themselves with bills nearly 40 percent higher than they had been a few months earlier.

As the state's energy companies and consumers struggled with the sticker shock of higher electricity prices, Davis began pursuing those responsible.

Ultimately, he directed blame at out-of-state energy producers and the Federal Energy Regulatory Commission (FERC).

As early as November 2000, California energy companies began to suspect out-of-state suppliers of manipulating the price of energy. A study by the Massachusetts Institute of Technology (MIT) found that energy providers made billions of dollars in excess profits in 2000 by cutting back power generation and creating artificial shortages.[7] Further revelations proved beyond a doubt that energy companies used California's low reserves as leverage for shutting down their power plants, thereby spiking energy prices.[8] Davis suspected as much in 2000, but the lack of public documentation and convoluted methods of energy trading made it almost impossible to detect criminal activity until 2 years later.

As California struggled to get control of mounting energy costs, Davis turned to the FERC for relief. He asked the Republican-controlled commission to investigate suspected wrongdoings and require energy companies to return as much as $8.9 billion in overcharges. The commission refused and blamed state leaders for the energy shortages. Meanwhile, Vice President Dick Cheney, formerly the CEO of a major energy company, criticized California's environmental regulations as the primary cause of the state's inability to provide enough electricity.[9]

Ultimately, Davis negotiated thirty-eight long-term contracts with several energy companies, including El Paso, Williams, Dynegy, and Calpine. Under the new terms, the companies agreed to provide power at rates well below the spiked prices that appeared intermittently during 2000 but considerably above 1999 prices. With growing claims of collusion, utility critics suggested that the state had now paid too much for the contracts.[10]

All of this took a toll on the popularity of Gray Davis and the policies of his administration. In December 2001, a statewide poll asked likely voters "How much of a problem is the cost, supply, and demand for electricity in California today?" In reply, 48 percent called it a "big problem," followed by 33 percent who classified the issue as "somewhat of a problem."[11] Even more telling, just as Californians were about to cast their ballots for governor in 2002, another survey asked which of the candidates would do a better job on electricity and energy policy. Responding voters, by a margin of 44 to 38 percent, said they preferred Republican challenger Bill Simon over incumbent Gray Davis on these issues.[12] Clearly Davis was in trouble, but he survived the election.

Whatever the realities concerning energy policy, voters in California seemed to have made up their minds as to culpability. Despite the lack of power plant construction well before his election, evidence of energy company manipulation, and a hands-off attitude from the national government, Gray Davis was assigned much of the blame.

A Plunging Economy

It is said that when things are going well throughout the nation, they are even better in California. Likewise, when things are going poorly throughout the nation, they are even worse in California. Nowhere is that more clearly evident than in the national economy. The recession that slowed the nation between 1990 and 1992 nearly crippled California; likewise, the recovery that benefited the nation throughout the rest of the decade was stellar in California. It is in that context that we examine the state's economy during the first few years of the new century. As with the rest of the nation, California's economic well-being soured, only more so.

Throughout the recession, California's unemployment rate stayed well above the national average. In fact, throughout 2002, the state's unemployment stood about a full point above the national average. As recall momentum picked up in 2003, the economy continued to languish, although the gap between the national average and California's narrowed a bit. Whereas the nation's unemployment hovered at about 6.1 percent, unemployment in California remained stuck between 6.6 and 6.8 percent. Other indices also showed that California's problems continued to exceed those of the nation. Personal income growth in California, for example, rose 2.2 percent during 2002, compared with the national average of 2.7 percent.

Manufacturing jobs led the economic decline. Between January 2001 and May 2003, the number of people employed nationwide in these typically high-paying jobs fell by 13.3 percent; California, however, lost 14 percent of its manufacturing jobs. March 2003 was a particularly brutal month for economic calamity in the rapidly tarnishing Golden State. According to the U.S. Bureau of Labor Statistics, employers laid off 109,000 employees during that month alone, a staggering 36,000 of whom were let go in California.[13] March also was the first month when the recall petition began to pick up steam. A coincidence? Maybe, maybe not. And during the month of July, when the recall qualified for the ballot, California lost 21,800 jobs, nearly half of the 44,000 lost nationwide.[14]

Some of the biggest problems occurred in sectors of particular value to California. Consider technology, which, over the past decade has accounted for more than one-quarter of the nation's exports but 52 percent of the exports from California. During 2001, growth in the state's technology segment slowed by 90 percent, leading to a reduction of 17 percent in the state's exports; in comparison, the national average fell by 6 percent.[15] Employment in the technology sector fell as well. Between January 2001 and December 2002, U.S. technology employment dropped by a half-million jobs, more than 100,000 of which were lost in California.[16] Clearly, one of the state's economic cornerstones was crumbling.

20

Ironically, employment during the entire Davis administration actually fared better than employment for the nation as a whole. From the time that Davis assumed the governorship in early 1999 until late September 2003, California actually added 658,000 nonfarm jobs—an increase of 4.8 percent. During the same period nationwide nonfarm employment rose by a paltry 1.9 percent, or less than half the rate of gain in California. But in 2003, voters weren't looking back to 1999 as much as they were wondering why the state unemployment rate remained well above the national average.

As 2003 neared its end, some signs suggested an economy on the verge of upward movement. In late September 2003, the well-respected quarterly *Anderson Forecast* at the University of California, Los Angeles (UCLA) predicted an upswing of 1 percent in state employment for 2004, the first increase in 3 years, along with even more robust growth in 2005. The same study pointed to increased venture capital spending and growing consumer confidence, along with anticipated increases in personal income.[17]

All that sounded hopeful to a state that had languished in the economic doldrums for so long. But for many, the promise of improvement was too little, too late, and these people put the blame squarely on the shoulders of Gray Davis. Indeed, a survey taken by The Field Poll just weeks before the October recall election found 83 percent of those interviewed agreeing that the state was in the midst of bad economic times, up from 56 percent in 2002—the worst assessment since 1993.[18] That concern became another element of the recall campaign.

The Budget Deficit

It's amazing how fast budget projections can change directions in government. One day the state is gasping for revenue, the next day it's flush with dough. And then another day after that (figuratively speaking), it's broke again. Such is life on the California highway, where economic traffic flows freely one moment and is jammed to a standstill the next. California is the state that, seemingly overnight, developed a revenue gap of $38 billion dollars. How did this happen?

It wasn't easy. After struggling with a miserable recession during the 1990s to make revenues equal to expenditures, California's future seemed to change. Suddenly, with the dawn of the new millennium, the Golden State once again seemed to glisten. In actuality, the years of great growth had stopped by 2000, when recession slowly began to emerge; state spending, however, continued at levels suggesting unbridled optimism.

There was a bit of delay between the economic slowdown and spending reassessments—not unusual, given the way businesses organize their tax payments on a quarterly schedule and individuals who don't settle their tax bills until April of the next year. In January 2000, the California Department of Finance estimated that the state would collect $3 billion more than the governor projected for the coming fiscal year, beginning in July 2000. Then-Speaker of the Assembly Antonio Villaraigosa predicted that the newly found funds would go to public education, health care, and infrastructure—areas that had languished for the preceding two decades.[19] In fact, the state was already living on borrowed time.

The state budget bloomed. The single most targeted area was public education. During the first Davis administration, state spending on K–12 public education accounted for 36 percent of the additional money flowing into the treasury; higher education took another 13 percent of the growth.[20] Some of the additional funds were used to pay for the 7 percent growth of K–12 enrollment. Most of the new money, however, was dedicated to digging the state's education programs out of a huge funding hole. In 1998, California ranked forty-third of the fifty states in per capita spending on public education; by 2003, the state had climbed to a more respectable twenty-sixth place. These infusions were reflected in improved student scores on scholastic aptitude tests (SATs).

Nearly half of the increased expenditures went to health and social services. Between 1998 and 2002, roughly paralleling the first Davis administration, the state embarked on a massive program for children of families who were without health insurance. By 2002, more than 660,000 children were signed up in a program called Healthy Families at a cost to the state of about $500 million. Davis supporters defended the program as an inexpensive way of avoiding major medical issues down the road; conservatives identified Healthy Families as an example of a state budget out of control.

Then there was the matter of prisons, which accounted for 25 percent of the additional funds garnered during the first Davis administration. Much of the new spending was mandated either by new federal government rules or by initiatives enacted by the voters. But it was the new contract with prison guards that seemed to get up the dander of conservative opponents. With the state deficit well publicized in 2002, the Davis administration negotiated a whopping 35 percent pay increase over 3 years with prison guards, whose union just happened to be one of the largest contributors to the Davis reelection effort. This relationship became a prime example of the "pay-to-play" misconduct cited by Davis opponents.

Meanwhile, as expenditures grew, revenues declined. Before long, the state clearly was spending more than it was taking in—and the problem began a lot earlier than most people realized. As early as September 2001, state economists predicted a revenue shortfall of at least $1 billion during the second quarter of the 2001–2002 fiscal year, which would begin on July 1 and end on June 30. "My sense is that we're in for a tough three or four months here," the state's chief economist warned in October 2001.[21] By the end of the fiscal year, in fact, a modest $2.6 billion surplus had been wiped out, and 3 or 4 months stretched into 3 or 4 years. In January 2002, the state budget projected a shortfall for the upcoming 2002–2003 fiscal year of $12 billion. By May 2002, forecasters revised the shortfall to $23.6 billion, and the governor and legislature went into a serious budget-cutting mode, lopping more than $8 billion from future spending. Yet, even after the 2002–2003 budget was put into place, analysts saw a continuing hole of $21 billion. And by January 2003, Governor Davis cited revised projections that placed the deficit at $34.6 billion.

How did it happen? Part of the problem emerged from the state's need to cough up an unexpected $6 billion to cope with surging energy prices. The financial difficulty was exacerbated, no doubt, by the sudden need to direct more than $1 billion in state funds for homeland security after September 11, 2001. Both of these were totally unpredictable expenses, yet the bills had to be paid.

But by far the biggest source of the revenue shortfall lay in the dramatic decline of personal income tax dollars resulting from the state's suddenly shaky economy, particularly in relation to stock options and capital gains. In fact, revenues from the personal income tax, which accounts for about half of all general funds, dropped by 26 percent between 2001 and 2003. Revenues from state sales taxes, the other major source, accounting for about one-third of state revenues, remained flat during the same period.

As the national recession took hold and the high-tech sector in California particularly felt its impact, companies downsized and in many cases went out of business altogether. With those precipitous declines came an equally precipitous decline in state revenues and new demands on state social services. If a picture is worth a thousand words, the figures in Table 2.1 speak volumes on the extent to which the decline in income taxes from stock options and capital gains led the drop in state income.[22] As high-tech went into a tailspin, so did the state's budget. But despite their efforts to pare expenditures, the governor and legislature simply could not keep up with the fiscal free fall.

Resolving the budget gap was nearly impossible to do in Sacramento's increasingly partisan environment. Although Democrats had majorities in both houses of the legislature, their numbers weren't large enough to meet California's constitutional requirement of a two-thirds majority for approval of a budget without Republican assistance. Seeing an issue in the making, Republicans weren't about to help out. In June 2002, the two sides sparred over the budget for a full 2 months

Table 2.1

State Income from Stock Options and Capital Gains, 1995–2003

Year	Amount (in billions)
1995–96	$ 2.6
1996–97	$ 4.0
1997–98	$ 5.5
1998–99	$ 7.5
1999–00	$12.7
2000–01	$17.7
2001–02	$ 9.5
2002–03	$ 9.7

Source: Department of Finance, Office of the Governor.

into the *next* fiscal year, finally, on August 31—the latest date ever—passing a "balanced" budget built on questionable assumptions. The new budget foundation became the basis for further deterioration in 2003. Meanwhile, the more the governor and legislature pointed fingers at each other, the more the public fumed.

And Then the Rains Came . . .

Take a close look at California's political system and you wonder why the state hasn't already fallen apart. Elected offices are disconnected, policy makers have no opportunity to develop expertise, lawmaking is almost impossible, political parties are disorganized, and direct democracy has emerged as a substitute for the power normally accorded to elected officials. It all boils down to a dysfunctional political arrangement that promotes confusion while thwarting accountability. It is also a system where it's much easier for those in charge to point fingers at one another than to get anything accomplished. This condition is not necessarily their fault, mind you, given the lack of linkages among those policy makers who might get something done otherwise.

It is against this porous structural backdrop that Gray Davis faced the voters in 2003. True enough, Davis was not exactly the warm and fuzzy type; moreover, he was not afraid to move ahead on his own en route to reaching his goals, often angering other officials who believed they should have been consulted, not told. Nevertheless, Davis broke no laws as he struggled to rescue the state from its economic malaise. But governors historically get too much credit when times are good and too much blame when times are bad. In fact, we often count on them to do much more than they are constitutionally able to do[23]—an expectation that generally has no foundation other than hope.

When California's economy tanked, Gray Davis was in charge because he was governor. Likewise, when the state's power went off, Gray Davis was in charge. And when the deficit spun out of control, Gray Davis was in charge. Whether Davis could have actually fixed these problems by himself is not the question. Simply put, in a state where the public expected its highest elected official to manage the various formal and informal obstacles before him, he did not fulfill expectations. Further, whatever his concerns about these issues, Davis did not seem to relate them to the public in the compassionate, caring way that they expected and needed. For these reasons, Davis's biggest problem may have been in the public's perception of his personality rather than any realistic appraisal of his effort. And in results-oriented, personality-driven California, that lapse in judgment was more than enough to put the governor on the carpet—and send him away.

Part II

The Shaping of the Battle

3

The Legislature

Gridlock Defined

It's one of the first things you learn about government as a kid—legislatures make the laws. Whether it's at the national, state, or local level of power, we're taught early on that legislative bodies are the embodiment of the people. And why not? These units are collections of people with roots close to ours. Members of the other branches just seem removed from our daily lives, but legislators are people we know. They're the folks who went to our schools, attended our religious institutions, and worked in our communities before moving on to their elective offices.

Whatever the concerns with legislative institutions, individual lawmakers usually enjoy tremendous popularity within their districts. Congress may screw up, but our representative is a good person. The city council may be in the dark on key issues, but the member elected from our district listens to us, thank you. Never mind that chief executives and courts, for that matter, often have greater impact on the political process than legislatures; we identify most with our lawmakers. And so it goes with California's legislature: they're the group we count on to make sense of things when things just don't make sense otherwise. Simply put, we trust our local legislators not to let us down.

The problem is that modern state legislatures can't compete with their counterparts in the executive and judicial branches. Lacking the bureaucratic command enjoyed by the executive and missing the ultimate judicial power to declare policies unconstitutional, legislatures have a tough time balancing the other two branches of government.[1] More often than not, they're a step or two behind, rather than in front of, the policy-making curve.

The imbalance is even greater in California, where the legislature operates with three major limitations. Two of them have been imposed upon lawmakers from without: First, the state Constitution requires an extremely high vote to pass meaningful legislation. Second, the voters have placed strict limitations on how long legislators may serve. In addition to these two con-

29

straints, the legislature has perverted its redistricting procedure, the process through which the state's population is divided into equal-sized districts every 10 years. Combined, these three elements have both weakened the legislature as an institution and left it caught in increasingly partisan struggles.

All of this occurs in an environment where upwards of 5,000 bills are introduced in the course of a 2-year cycle—many of which will receive multiple hearings by several legislative committees prior to any final votes on the floor. Simply put, the legislature operates in a political zoo without a zookeeper. As a result, this branch of state government has failed to rise to the occasion as a coequal branch ready to assume policy-making obligations. That inability to resolve conflict became a contributing factor to the growth of public exasperation. If the recall of Gray Davis was an indictment of failed political leadership, then the legislature should have been added as an unindicted co-conspirator.

The Problem of the Two-Thirds Vote

Passing major legislation isn't easy in California. Thanks to the concerns of reformers when the document was overhauled in 1879, the state constitution requires an *absolute* two-thirds vote from the legislature to enact a budget. *Absolute* means two-thirds of the 80-member assembly and 40-member senate, or 54 and 27 votes respectively. Here's another way to look at the issue— votes of 53-to-0 and 26-to-0 fail when it comes to budget matters. Only two states have budget-passing requirements as tough as those in California; the other forty-seven have lower thresholds for passage.

The reformers included the absolute two-thirds requirement to make sure that the legislature would not be bowled over by a special interest group, particularly the Southern Pacific Railroad, which dominated state politics at the time. They succeeded; that's for sure. But on the way to assuring landslide votes, the reformers also gave us a legislature dependent on both major parties to get the necessary votes over an impossibly high political bar. And in case you haven't noticed, the Democrats and Republicans in the California state legislature don't have much use for each other.

Between 1978 and 2003, the legislature failed to pass an annual state budget by the June 15 constitutional deadline in 17 of the 25 years. That might be a good passing percentage for a football quarterback, but it revealed the legislature to be an unreliable part of the budgetary process. Each year, the Democrats ultimately were able to pry loose the necessary Republican votes after last-minute horse trading.[2] In the process, the governor's responsibility for organizing and shepherding the budget through the legislative process has grown.

30

Budget matters in the legislature became particularly sticky as the deficit grew in the early part of the current decade. Shortly after his November 2002 reelection, Governor Gray Davis predicted that the deficit through the end of the 2003–2004 fiscal year would reach $21 billion—a number that turned out to be just about half of the eventual $38 billion hole. In December 2002, Davis asked the legislature to make a start on addressing the problem by enacting $10.2 billion in midyear cuts, $3.4 billion for the rest of the 2002–2003 fiscal year and $6.8 billion for fiscal 2003–2004. The legislature met symbolically for a day on December 3 to swear in new members and then recessed until January 6, 2003. Either Governor Davis or the legislature's leaders could have called a special session to deal with the budget issue immediately; neither did so.

When the governor gave his State of the State address in early January 2003, he upped the ante. Citing new figures that placed the deficit at $35 billion, Davis proposed $8.3 billion in new high-end income taxes and a 1-cent increase in the state sales tax. In addition, he asked the legislature to slash state spending for the 2003–2004 fiscal year by $20 billion, impacting every program and service except for state prisons.

This time, Democrats were prepared to move forward, but with only 48 of the 54 votes needed in the assembly and 25 of the 27 votes needed in the senate, they were unable to act without Republican help. Republicans, however, refused to budge, stating that they would not consider 1 cent in new taxes without Democrats agreeing to a permanent constitutional cap on state spending and repeal of several labor and environmental laws.[3] In late February 2003, Republican resolve stiffened further, as Senate Minority Leader Jim Brulte proposed a budget without any new taxes and cuts of more than $25 billion, or about one-third of the state's general fund.[4] Democrats bridled at the thought and accused the Republicans of heartlessness.

The stalemate continued to and through the June 15 constitutional deadline, although the deficit eventually rose to $38 billion. Democrats lacked the necessary complement to pass a budget; Republicans remained dead set against any taxes. The only change came through the addition of $4 billion worth of motor vehicle taxes, which were reinstated by the governor because of a provision in the original law allowing the governor to restore the tax to earlier levels if the budget went into the red. Republicans reeled at the thought, but the Democrats had sufficient votes to prevent a vote blocking the governor's action.

By May, new figures provided by the state department of finance pegged the deficit at $38 billion. Davis argued further for higher personal income and cigarette taxes, along with $10 billion in loans, which would be repaid from higher sales taxes. The Republicans dug in their heels—so much so that in June, Senate Minority Leader Brulte threatened to personally campaign against any Republican who voted to raise taxes as part of the budget solution.[5] Finally, on July 30, nearly 1 month into the new fiscal year, the legislature passed a Republican-endorsed budget. The document included $7 billion in cuts and $14 billion in loans; it left an $8 billion hole for 2003–2004. New fees, including restoration of $4 billion in unpopular motor vehicle registration charges, made up the rest of the difference. Nevertheless, the package was without any tax increase, signaling a major victory for Republicans. Clearly the days of the Democrats picking off the necessary Republican votes to enact budgets were over.

Despite obvious partisan bickering, the delay was not all the legislature's fault. A statewide poll taken in late May 2003 found that the public was dead set against any cuts in key programs. As if to make matters worse, the same poll found large majorities opposed to increasing all major taxes except those on cigarettes (see Figure 3.1). Given the strong will of the public and lack of leadership in the legislature to either raise taxes or drastically cut expenditures, the budget process remained stalled.

The irony is that even though the voters were paralyzed, they expected the legislature to solve their dilemma. A survey by The Field Poll taken in the midst of the 2003 budget standoff found the electorate blaming the legislature almost as much as the governor for the budget deficit[6]; on the other hand, according to another Field Poll conducted in September 2003, the public was overwhelmingly in favor of staying with the absolute two-thirds vote for budget matters, despite the costly standoff.[7] Such was the environment in which the legislature had to conduct its business. And it was taking its toll on Governor Gray Davis.

The Term-Limits Paradox

Passed easily by the voters as an initiative (Proposition 140) in 1990, term limits have become sacred in California. The idea behind term limits is simple enough—legislators should not stay around Sacramento too long lest they become too comfortable in their positions of power and abuse the public trust. Thus, since 1992, members of the California State Assembly have been limited to three 2-year terms, while senators have been limited to two 4-year terms.

As a result of term limits, experience is at a premium in the legislature. The position of speaker of the assembly, once a source of both power and organization, is now turned over from one legislator to another on a regular basis. Typically, a member is elected to the speakership in his third year of service and rotated out during his fifth year. Speakers usually use their sixth year to

Figure 3.1 **Public Attitudes on Taxing and Spending**

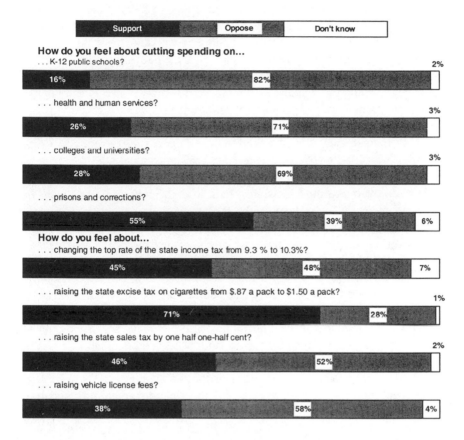

Results based on telephone interviews of 2,003 adult California residents conducted May 22 to June 1, 2003. The sampling error is plus or minus 2 percent.

Source: California State Budget Survey, a collaborative effort of the Public Policy Institute of California and the James Irvine Foundation.

position themselves for election to the state senate or another elected post if possible. Individuals elected to the position of senate president pro tem, the top job in the senate, last a bit longer, but usually with just as little clout.

If it seems that this system prevents anyone from being very powerful, it's because that is the case. But at the same time, without any institutional memory or leadership, the legislature has no guidance. As a result, legislators are not only confused about the process but depend on others to help them through it.

Few understand the workings of the legislature as well as State Senator

John Vasconcellos, who was first elected to the assembly in 1966 and who will become one of the last victims of term limits in 2006. Focusing on the massive inexperience of members, Vasconcellos recently said: "A third of the folks have been here for six months and they walk into a $38 billion hole. There is no mentoring. No seasoning. No history. No loyalty. This is an awesome responsibility. Not just a trite piece of work. You really have to understand what's going on."[8] State Senator Debra Bowen, who freely admits that women and minorities have benefited from term-limit turnovers, can't imagine what it would be like for the model to be used in the private sector: "I ask people in industry, 'How well would your company run if every two years you got a new CEO and one-third of your board of directors left?'"[9]

All this has come to a frightening climax with the budget morass. When the state enjoyed a surplus, legislators had little difficulty in finding ways to spend money or cut taxes. But once the deficits hit, their inexperience showed. By 2003, they were simply overwhelmed by their responsibilities. Bills were turned inside out, with completely new language, on a moment's notice; hearings were held late into the night, making public participation all but impossible; and legislative rules were often ignored in the name of expediency.

Then there are the lobbyists. It's no secret that lobbyists curry favor of legislators via campaign contributions. That relationship—like it or not—is relatively transparent. But beyond money is the simple fact that lobbyists possess enormous amounts of information that legislators either don't have or don't know how to get. As Dan Walters and Jay Michael note, when it comes to management of their issues in the legislative process, lobbyists have "spent months, even years, gathering information, attending legislative hearings, [and] negotiating with each other."[10] How can term-limited legislators who are in the same vicinity for a fraction of the time have nearly as much expertise? Out of this imbalance, legislators lean on lobbyists for their one-sided knowledge as much as anything else. No wonder that so many bills tied to special interests get through the legislative process.

Finally, there's the imbalance between term-limited legislators and the governor. "But wait," one might ask. "Isn't the governor term-limited, too?" That's true enough. But the difference between the governor and the legislature is that the governor has access to a large bureaucracy that most legislators don't even know how to mine. As a result, the legislature has great difficulty questioning assumptions and positions of the executive branch. When Davis spit out study after study regarding the budget crisis along with his recommendations, legislators were hard pressed to dispute them with equally powerful data. And when they did, lawmakers had great difficulty attracting the kind of press the governor could draw. Thus, their only recourse was to do nothing, to hold back and refrain from taking positions, just to accommodate the governor. That might have given legislators some short-term satisfaction, but in the end, their inability to find budget solutions was seen by the electorate as another sign of the governor's inability to manage the legislative process.

The inherent weakness of term limits is that it leaves the legislature unable to compete with interest groups, the bureaucracy, or the executive branch. Rather than frame policies, legislators are forced to backpedal just to keep competitors at bay. As a result, they have become defensive players in the political process. Collectively, these undesirable traits became part of the backdrop in the recall movement of 2003.

Safe Districts and Dangerous Legislative Conflicts

These days, students of the California legislature hear the same questions over and over again. "What's happened to moderation and compromise in the state legislature? Why does there have to be such acrimony?" Why are relations between the members different today than in the past? The answer is that these values left the day that safe legislative districts arrived.

Blame it on something known as redistricting. Every 10 years, right after the census, states—if their populations have changed—are assigned their new numbers of members of the U.S. House of Representatives. The state must then be carved up into new districts of similar population sizes. After the 2000 census, California went from 52 members in the House to 53 members, and the legislature organized congressional seats into evenly populated districts. At the same time, the legislature redistricted assembly and senate seats, according to the requirement established almost 40 years earlier in a landmark U.S. Supreme Court decision.[11] Even though their numbers remained at 80 and 40 respectively, some areas of the state had gained or lost population relative to others. Thus, the legislature was redistricted.

From the start, Democrats and Republicans agreed to draw district lines in such a way that the status quo would prevail. The plan, agreed to by wide margins in the normally divided legislature, left the legislature with virtually the same numbers of Democrats and Republicans after redistricting as before. Just prior to the 2001 formula, assembly Democrats outnumbered Republicans by a margin of 50 to 30; after the 2002 elections, Democrats had a 48-to-32 edge. Election outcomes were even closer in the senate, where, before the redistricting plan, Democrats outnumbered Republicans by a margin of 26 to 14; after the 2002 elections, Democrats prevailed over Republicans by a margin of 25 to 15. The result: A strangely bipartisan plan "designed to reduce competition at the ballot box and preserve the partisan status quo in the Legislature," according to one account at the time.[12]

There were important political reasons to cut the deal. If they tried to leverage more seats with creative redistricting, Democrats feared a court suit similar to the one that was ultimately settled by the California Supreme Court in 1991, when district lines were drawn by a commission appointed by the governor, not by legislators.[13] On the other side of the legislative aisle, Republicans worried that fighting a Democratic plan would further diminish their already fragile positions. Thus, the two sides reached accommodation on this normally touchy topic in record time. Governor Davis signed the bill almost as soon as it reached his desk.

The agreement produced astonishing results. Rather than enhancing competition within districts, the plan made a mockery of competition. Instead of districts where candidates of either major party had a chance of winning, most districts were packed with majorities of one party or another—guaranteeing victory for the candidate of that party. After reviewing the district-by-district voter registration breakdowns, the *California Journal*, a well-respected observer of California politics, suggested that only 3 of the 80 assembly districts and 1 of the 20 state senate districts would be close to competitive in the 2002 election.[14]

What does all this have to do with legislative stalemate? With virtually every legislator hailing from a district made up of overwhelming registration majorities of his or her own political party, the elected individual doesn't have to worry about serious opposition at the general election in November. All the real contests were in the primary elections, when each party picked its nominee. In a heavily Democratic district, the "most Democratic" candidate—usually the most liberal—is most likely to win. In a heavily Republican district, the "most Republican" candidate—usually the most conservative—generally wins. Moderates of either party were less likely to win nomination. Consequently, the gap between the parties in the legislature grew, with few moderates in the middle to bridge that gap. Legislators feel free to "stand on principle," which may sound good from the standpoint of staying true to one's beliefs, but it's rotten in terms of finding common ground for legislation and other policy decisions. These circumstances created a highly polarized legislature consisting of individuals with little desire to pursue compromise over principles, and the public was the loser for it.

Adding Fuel to the Fire

Was the gridlock in the state legislature responsible for the recall of Governor Davis? Absolutely not. Was the growing paralysis regarding management of the state budget year after year a contributing factor to the governor's demise? Yes, without a doubt. Between a chief executive who was hard to work with under the best of conditions and the organizational disarray that produced polarized stubbornness, the legislature was not a fun place to be in 2002 and 2003. With every report about a soaring deficit, the collective funk of the state's lawmakers seemed to grow. The angst was manifest in the most overdue budget in state history in 2002, followed by another overdue budget document in July 2003.

True enough, none of this was directed specifically at Gray Davis or caused by the governor. Then again, the lack of cooperative spirit didn't help the governor either. Most of all, for an electorate increasingly concerned about a state in disrepair, the lines separating the executive and legislative branches didn't really matter. If the legislature was unresponsive to state needs, voters seemed to be saying, it was because the governor was not leading them to the Promised Land. In the end, virtually all responsibility for California's dysfunctional governmental system seemed to land at the feet of Gray Davis, fairly or not.

4

Governor Gray Davis

Triumph, Decline, Denial

The power crisis, the recession, and the budget deficit were crucial factors in the shaping of the perfect storm. But other states have faced and are facing these same problems or worse, and their governors have not been recalled (yet). Other factors also contributed to the recall, including the governor's response to these big problems, his personality and political style, and, ultimately, the availability of alternatives to the sitting governor.

The combination was nothing short of lethal. After all, other executives—such as Bill Clinton—have gotten into trouble but survived because they were seen to be doing their jobs to the satisfaction of a majority of the people and because they had a reservoir of popularity as well as a network of strong political allies. Gray Davis, uniquely for someone who had risen to such a high position, had none of these.

Who Is Gray Davis?

Joseph Graham Davis, Jr., was born in New York City on December 6, 1942—a pre–baby boomer. He got the nickname "Gray" from his mother, although many say it suits his personality. He spent his childhood in upscale Greenwich, Connecticut. The family moved to Westwood in Los Angeles when Gray was 11 years old. His father, who sold ads for Time, Inc., drank and drove the family into debt and finally abandoned Gray's mother and her five kids. Gray went on to Stanford University, where he worked to put himself through school and help out his family. He also played varsity golf and joined the Reserve Officers' Training Corps (ROTC). He went on to law school at Columbia University in New York City. Military service—an obligation of his participation in ROTC—followed. U.S. Army Captain Gray Davis served a year in the Signal Corps in Vietnam and was awarded a Bronze Star.[1]

"I was pretty naive growing up," Davis told a reporter for the *San Jose Mercury News*, "and I just thought everyone did their duty. America was fighting a war, everyone kind of had to share the burden. The burden of this war fell disproportionately on minorities and whites that were less well educated. That's why I'm a Democrat."[2]

Davis got into politics by working on Tom Bradley's campaign for mayor of Los Angeles in 1973 and Jerry Brown's campaign for governor in 1974. When Brown was elected, Davis became his chief of staff, serving in that office until 1981. Given his reserved personality, the behind-the-scenes role of chief of staff might have suited Davis better than the out-front role of governor. While working for Brown, Davis met and married Sharon Ryer, a flight attendant he got to know on trips between Los Angeles and Sacramento. It's said that she took offense when he arrived late for their first flight together and failed to apologize.[3] Later, she would assist her husband by giving his otherwise stiff public persona a softer, more human side.

Davis went back to L.A. and ran for state assembly in 1982, winning a seat that included Beverly Hills—perhaps not the most representative cross section of California Democrats. In 1986, he ran for state controller, an office responsible for overseeing the state's taxing and spending. Even in these early offices, issues arose about Davis's fund-raising practices, although no criminal wrongdoing was proven.

In 1992, Davis ran for the Democratic nomination for U.S. Senate, losing to Dianne Feinstein in a nasty campaign. Davis compared Feinstein to Leona Helmsley, New York's "Queen of Mean," because she had been fined for violating campaign finance laws (not unlike Davis himself). "Helmsley blames her servants for the felony," the ad declared. "Feinstein blames her staff for the lawsuit. Helmsley's in jail. Feinstein wants to be governor."[4] The attack hurt Davis more than Feinstein as Democrats turned on him. But Davis is nothing if not tenacious. In 1994 he ran for lieutenant governor—not the most high-profile of positions—and won, even as Republican Pete Wilson was winning reelection as governor.

With Wilson facing term limits in 1998, Davis was well positioned to run for governor. He launched his campaign early, raising money with vigor and presenting himself as the "experienced" candidate based on his long career in public office. Davis won a tough race for the Democratic gubernatorial nomination against businessman Al Checchi and U.S. Representative Jane Harman, both of whom drew heavily on their personal wealth to fund their campaigns. Checchi alone spent $38 million. Labor and other allies helped Davis stay competitive, but Davis also benefited when Checchi ran ads attacking Harman early in the race. He damaged her chances, but instead of rallying to Checchi, voters turned to Davis, who minimized his own negative campaigning. Davis, who had been written off by some observers early in the election year, pulled off a surprise victory.

In the November 1998 election, Davis faced the state's attorney general, Republican Dan Lungren, a staunch conservative. Davis's ads reminded voters that conservative Lungren was out of step with California voters on most social issues while presenting Davis as the moderate, tough-on-crime candidate. While Lungren was far more articulate in debates, his lackluster campaign could not overcome his conservative label to reach beyond his core constituency. Davis beat Lungren by 20 percentage points, a virtual landslide.

But was it a mandate? Throughout his political career, Davis seems to have won races almost by default rather from the overt enthusiasm of his supporters and the voters. He chose his races carefully and timed his advance perfectly. By the time he ran for governor, he was the senior office-holder in his party and could claim the nomination almost by right. Even so he had to fight for it. There again, his luck held when his main opponents canceled one another out and the Republicans nominated a weak candidate. He wasn't necessarily a "lesser evil" choice for voters, but he never evoked passion or enthusiasm. His was a plodding rise to the top that concluded with a sweeping victory. But in retrospect, that victory may have been hollow. He seems always to have been the marginally acceptable candidate rather than the candidate of choice.

Perhaps personality has something to do with the ultimate shallowness of his support. Observers agree that Davis is reserved, even cold. After 30 years in public life, people still asked "Who is Gray Davis?" When Jerry Brown was governor, we knew about his beat-up state car, his small apartment, and even whom he was dating. Even his uncharismatic successors George Deukmejian and Pete Wilson had more public personae than Gray Davis. All that people seem to have known about Davis's personal life was that he had a wife whose more outgoing personality warmed the public view of her husband—but not enough. Does he have a dog like most presidents? Does he play golf? Does he like movies? Would we see him at the Oscars? Or a baseball game? (During the recall campaign, we did learn that he eats turkey sandwiches every day for lunch and relaxes at night by watching ESPN cable television, and that actress Cybill Shepherd thought he was "a good kisser" when they made out more than 30 years ago.)

Some of his neighbors in Sacramento seem to have actively disliked him. You "never see him out there," one told the *New York Times*. "He just goes in that house and shuts the doors and closes the curtains."[5] And when he faced the ultimate challenge—the recall—he seemed to have few friends or close political allies. He has "zero personal relationships," San Francisco Mayor Willie

Brown told an interviewer. "He's one of the more self-centered politicians I've met in my life," said Delaine Eastin, a Democrat and the former state superintendent of public instruction. "It was all about him. What would get him good press? What would position him with his donors?"[6] But if his personality caused political problems for Gray Davis, so did his style of governing.

Governor Davis: "Implement My Vision"

Perhaps the seeds of the recall movement were sown in 1999, shortly after Gray Davis and Cruz Bustamante assumed their respective offices of governor and lieutenant governor. That the two seasoned politicians were both Democrats did not necessarily mean they had run for office as a team. To the contrary, California electoral law permits people to pursue executive branch offices individually, rather than via a party ticket that couples political party nominees for president and vice president. Thus, candidates of the same political party often have little in common and even less loyalty to one another. In the past, that lack of connectivity has sparked confrontations between the state's two top leaders, especially when the governor is out of state and the otherwise powerless lieutenant governor gets to play grownup as acting governor for a day or two.[7]

As governor, Gray Davis wasted little time in attempting to take control of the state's massive, yet poorly organized and often overlapping governing structure. But rather than carefully build enduring political alliances, he charged ahead single-mindedly. His first public confrontation was with Bustamante. The two leaders had barely settled into their offices across the hall from each other when Davis, largely the political centrist, appeared to waffle over Proposition 187, a divisive anti–illegal immigration ballot initiative passed by the voters in 1994 that was ultimately severely limited by the federal courts. As the case against Proposition 187 was going through the judicial process, Bustamante publicly chided Davis for not having moved sooner to drop the state's appeal of the federal court ruling that overturned the ballot measure. Davis responded by taking away most of the capitol parking passes for Bustamante's staff.

There are others, too, who found themselves in Davis's way. In 1999, after tangling with the legislature on a series of thorny budget issues that he deemed too costly, Davis complained that individual legislators just didn't have the broad perspective of a statewide elected official such as himself. The job of the state legislature, he declared, was "to implement my vision."[8] A year later, the governor said much the same about the courts, when he proclaimed that those he named to the state bench should reflect his values on positions of importance to the governor.[9]

On each occasion, the governor backpedaled in the days that followed, but his strident positions took a toll nonetheless. Davis's actions also laid the groundwork for an absence of loyalty in the years to come—loyalty that could have come in handy as the recall movement picked up steam. In a state where personal relationships often compensate for gaps in the formal power structure, the governor seemed to be using up his share of goodwill early and carelessly. It wasn't long before the words "Davis" and "arrogance" were often used in the same sentence.[10]

Anyone who has sat through an elementary course in American government knows that the "separation of powers" concept is a hallmark of our political system. The idea is that each branch—executive, legislative, and judicial—has its own area of authority, and that policy can't be made without cooperation among the three. It's an awkward setup that creates a rather high threshold for getting things done. On the one hand, it's easy for one branch to gum up the works (and hopes) of the other two. On the other hand, if a proposed policy gets the nod from all three branches, there's a good chance that the public will be satisfied as well.

But "separation of powers" has been taken to extremes in California. Not only does the state mimic the federal government by assigning different responsibilities to the three different branches, but there are distinct separations of power and authority *within* the executive branch as well. Each statewide executive branch officeholder is not only elected independently of the others but also has responsibilities apart from the others. As a result, one statewide officeholder may pursue objectives that openly conflict with those of another.

The implications of this disjointed arrangement are profound. While the governor may be the officeholder with more responsibilities and power than the others, he does not necessarily have the ability to control the activities of those operating in other elected executive branch positions. To the casual observer, it appears that the California governor can't even control those who work *for* him. In fact, the governor sometimes can't control the activities of those who work *next* to him. Thus, in 1987, for example, then-Attorney General John Van de Kamp refused to represent Republican Governor George Deukmejian in a lawsuit over the governor's management of a "toxics" initiative, even though the attorney general is supposed to be the governor's lawyer.[11] The result: mixed messages from the executive branch about public policy issues and confusion for the public.

Similar confusion occurred with the administration of Gray Davis. Take the issue of Indian casinos. Gray Davis was lukewarm to the 1998 ballot initiative that legalized Indian gambling in California, and he has had testy relations with key Indian gaming interests ever since.[12] In fact, when the state found itself with an unprecedented $38 billion deficit in 2003, Davis proposed that the unusually low state taxes paid by Indian casinos be increased by an additional $1.5 billion. Tribal leaders wasted little time in "dissing" the proposal.[13] Meanwhile, Lieutenant Governor Cruz Bustamante has supported the concept and expansion of Indian gaming in California without new taxes—an attitude that gaming interests rewarded generously when Bustamante ran for governor as a replacement candidate in the recall election.[14]

So where does the state stand on controlling Indian gaming? It depends on whom you ask. But the average voter doesn't see *different* messages as much as *inconsistent* messages. That kind of assessment helped fuel the general conclusion that Gray Davis was erratic at best, two-faced at worst. And when you don't enjoy a reservoir of support from either the public or elected officials, such impressions can be damaging.

But while Davis had problems with other members of the executive branch—even though they were Democrats—his relations with the state legislature were worse. Despite having been a legislator himself, he never developed a good working relationship with the legislature or even with its Democratic majority, which should have been on his side. His "implement my vision" declaration set the tone, and things never improved.

Part of his challenge was that although Davis was a Democrat, he had been elected as a moderate, with conservative positions on capital punishment, the three strikes rule, and other law-and-order issues. Nor was he as liberal on a wide range of issues as his Democratic colleagues in the legislature. Many of the latter had been elected by majorities even greater than his in districts that were drawn so as to be strongly Democratic. After 16 years of Republican governors, they expected to be able to enact a wide range of liberal policies and proceeded to do so. The Democratic legislature went to work, passing a backlog of liberal bills, only to face vetoes from their own Democratic governor. Between 1999 and 2002, Davis vetoed 19.2 percent of the bills sent to him—a rate higher than that of his Republican predecessors, who also faced Democratic majorities in the legislature.[15]

Dan Walters, the dean of California's political columnists, described Davis as "a centrist triangulator during his first three years in his governorship, interposing himself as the decisive factor between business and professional groups and an increasingly liberal Legislature."[16] Davis disappointed key constituencies and supporters, from labor to Latinos and environmentalists, while granting tax breaks for corporations and raising money from business for his reelection campaign. He soon gained a reputation as a "pay-to-play" governor: campaign contributions seemed essential to get favorable consideration from the Davis administration. The California Teachers Association was asked for $1 million and the California Medical Association was told that campaign contributions were expected. Newspapers reported contracts, appointments, and policies that appeared to be linked to other campaign contributions. Claiming he was worried about a challenge from another wealthy candidate, Davis ultimately raised $78 million for his 2002 reelection campaign—a record for an individual candidate.

Davis's approval ratings in public opinion polls declined precipitously when the energy crisis hit California in 2001 and his response was perceived as inadequate. With the election approaching in 2002, the governor sought to shore up his support by moving to the left. He placated unions by signing workers' compensation legislation he had previously vetoed. He embraced environmental legislation to combat global warming. He signed into laws bills to extend gay rights and to make it easier for unions to organize farm workers. In short, he ceased to "triangulate" and attempted to rally the liberal troops around the Democratic cause.

These actions may have ensured his reelection, but, as in the past, Davis also benefited from a weak opponent in 2002. The leading contenders for the Republican gubernatorial nomination were Los Angeles Mayor Richard Riordan, a moderate and a former businessman, and Bill Simon, another businessman. Both Riordan and Simon are multimillionaires who were pre-pared to use their own money to fund their campaigns—exactly the kind of candidate Davis feared and attempted to anticipate with his relentless fund raising. In early polls, Riordan, the moderate, posed a greater threat to Davis in the general election than the more conservative Simon. Riordan also held the lead for the Republican Party nomination.

Although it's unusual for candidates of one party to run ads attacking those of another party during the primaries, strategists for Davis, fearing Riordan more than Simon, decided to do so, spending $9 million to tarnish Riordan's benign image. Having dredged up an old videotape of Riordan saying that abortion is murder, the ad pointed out that in 2002 he claimed to be pro-choice. Another ad said he opposed the death penalty while Davis supported it. At the time, Riordan's lead in the race for the Republican nomi-nation was so great that the Davis campaign's main intent may have been to weaken his support among moderate voters in anticipation of the November general election. But in the meantime, they put Riordan in the position of defending his moderate positions in a Republican primary—where conser-vative voters are crucial. It was almost a "have you stopped beating your wife?" sort of attack, with no defense that would not alienate some voters. Riordan responded ineptly and the ads hurt him with Republican voters.

Conservatives rallied to Bill Simon, who won the primary handily. Politi-cal mythology attributes Riordan's defeat to Davis's attack ads, but Riordan also contributed to his own loss. He proved to be an inarticulate candidate and a lackluster campaigner. Riordan himself asserts that Simon "buried me with ads saying [Riordan was] 'Embarrassed to be a Republican.'"[17] Ultimately, Simon's victory was an affirmation of the conservative dominance of the California Republican Party. It's been a long time since they've nominated a moderate for statewide office—which is one reason Republicans have not held a statewide office after the 2002 election.

Although multimillionaire Bill Simon "loaned" his campaign $10 million, he couldn't match Davis's huge war chest, nor could he redefine himself as a moderate after winning the Republican nomination on his conservative credentials. And the Davis campaign spent a lot of money reminding California voters how out of step with them Simon was on social issues—as well as running attack ads alleging that Simon, running on his know-how as a businessman, had mismanaged his own businesses and charities. Davis held a solid lead in the polls throughout the campaign and won in the November 2002 election.

But even a cursory analysis of that election made Davis's vulnerability apparent. In 1998, he won 4,860,702 votes—58 percent of those voting—to win by a 20 percent margin. In 2002, he won 3,533,490 votes—47 percent—winning by just 5 percent over a weak opponent. In 1998, some 96 percent of the voters chose the candidate of either the Democratic or Republican Party. In 2002, over 10 percent chose candidates of the minor parties—and 3.3 percent of those who went to the polls voted for no gubernatorial candidate at all. Democratic candidates for lieutenant governor, attorney general, and even superintendent of public instruction all got more votes than Gray Davis, the candidate at the top of their ticket. They ran stronger than Davis among core Democratic constituencies and in the increasingly important Central Valley. Davis's support among Latinos, for example, declined from 78 percent in 1998 to 65 percent in 2002[18]—aided by the refusal of the legislature's powerful Latino caucus to endorse Davis because he had not been sufficiently responsive to their concerns. In short, while Davis should have been gliding to a second-term triumph that could have made him a candidate for the U.S. presidency, his victory showed weakness more than strength.

Slow Reaction: A Governor in Denial

Sitting in the governor's office in 2002, Gray Davis could be satisfied with his record of accomplishment. In his first term, he had made education his top priority: he had succeeded in increasing per pupil spending by over $1,000 and in setting new standards by which academic success was to be measured. California ranked forty-third among the fifty states on per capita spending on public education when Davis took office but moved to twenty-sixth during his tenure. Financial aid for college students was expanded and tuition held steady during his first term. Nearly 700,000 children gained health care coverage and reforms of health maintenance organizations (HMOs) were introduced. Paid family leave was enacted for new parents or people caring for ill family members. Regulation of guns was expanded. The minimum wage was raised and unemployment and workers' compensation benefits were increased. The 8-hour workday was restored. A domestic partners registry was created and the rights of partners were expanded. Environmental progress included phasing out MTBE (a gasoline additive), forest and wetlands preservation, and stricter gas emission standards for cars and trucks.[19] Meanwhile, Davis stayed tough (some said too tough) on crime, taking a hard line on parole for anyone convicted of murder and continuing his support for prison guards and prison building. California—and the Davis administration—led the nation and the world on many of these issues and others as well. Many of these achievements, however, came from the legislature.[20] Davis signed the bills but initiated few of them.

Nevertheless, it was a record that might have merited a reelection landslide, except that Davis's cautious and introverted style made the accomplishments seem less than they were. In his first years, he seemed to follow public opinion polls and focus groups, despite his huge 1998 election victory. Dan Walters describes his style as "micromanaging, risk-averse, ever-rationalizing"—a nitpicker rather than a leader.[21] Others have called his style "incrementalist" and "transactional," emphasizing his preference for gradual rather than substantial change and the appearance of linking support for his campaigns to his approval of policies. Moreover, many of his most substantial accomplishments came as he faced reelection and sought to shore up support from traditional Democratic constituencies, sometimes by signing legislation he had previously vetoed.

In the end, all the accomplishments paled in the face of two major crises: energy and the budget deficit. When the energy crisis hit in 2001, the governor's approval ratings in public opinion polls plummeted from a high of 66 percent to 46 percent while his disapproval number almost doubled (see Table 4.1).

Although the governor eventually acted to address the crisis, his actions were widely perceived to be both late and costly. He seemed to be scrambling to fix a broken system at any cost rather than providing the strong, proactive leadership the public wanted. Meanwhile, a national recession took a grip on California. The terrorist acts of September 11, 2001, had their own impact on the economy, not least through a precipitous decline in tourism, a major industry for California. And back home in California, after having

Table 4.1

Approval Ratings of Governor Gray Davis (likely voters only)

	Approve	Disapprove	Don't know
September 2000	66	27	7
October 2000	61	30	9
January 2001	62	28	10
December 2001	46	48	6
January 2002	46	49	5
October 2002	45	52	3
February 2003	24	72	4
June 2003	21	75	4
July 2003	22	72	6
August 2003	25	72	3

Source: Public Policy Institute of California, "State of the Golden State," August 2003 (www.ppic.org).

floated the state's economy and tax revenues to their highest levels, the high-tech/dot-com economic bubble burst. During the boom, the governor and legislature had happily spent revenues and cut taxes—providing something for every political persuasion but failing to consider that the boom might end. When it did, both the governor and legislature remained in denial, avoiding hard decisions in 2001 and 2002 that might have made later years less painful. Republicans tried to raise the budget crisis in the 2002 election but gained little traction with the issue. Davis and the Democrats sought to divert voter attention to social issues, but beyond that, voters may have been in denial as much as their political leaders were. Spending and cutting taxes is so much more palatable than cutting programs and raising taxes, whether you're inside or outside of government.

The first decline in Davis's approval ratings came when the energy crisis hit. He held his ratings steady through the 2002 election and minimized discussion of the scale of California's budget problem. But shortly after the election, he announced California's $34.6 billion deficit, and in his January State of the State address attempted to deal with it. His approval ratings dropped to 24 percent and stayed there—or sank lower—as recall petitions circulated. Governor Davis proposed a budget that was balanced by spending cuts and tax increases with no borrowing; but as with the energy crisis, he seemed unable or unwilling to take an active leadership role, waiting instead for legislative compromise. By the time the budget was due (though not yet approved) in June, his approval ratings had hit the lowest point in the history of polling on California governors.

Davis Agonistes

When California ran into trouble, voters blamed politicians, especially the governor. While the state's problems were by no means all the fault of Gray Davis, his response was inadequate if not incompetent, and his ability to communicate with or show empathy for his citizens was sorely lacking. In such times of trouble the public might have rallied around a more charismatic and empathetic leader, as, arguably, they did around Bill Clinton. The governor might have used the bully pulpit of his office, appealing to the public through the media. Political allies in the legislature and elsewhere might have come to his defense. Democratic legislators might have shown more willingness to accept budget cuts; friendly Republicans (admittedly an oxymoron) might have backed off a little on new taxes. In fact, there was room for such compromise for, despite California's reputation as a "high tax" state, it ranks nineteenth among the fifty states in total tax burden.[22] The governor did propose some modest tax increases but declined to fight for them or take his case to the public.

And the legislature did not come to his aid. Gripped by gridlock and rigid partisanship, the legislature was incapable of compromise. Democrats watched with exasperation as conditions worsened and Republicans held firm against any new taxes, perhaps taking some pleasure in the difficulties of the governor and the opposition. But ultimately the legislature was part of the problem, as we will see.

"My greatest strength is everyone underestimates me," said the governor in response to his greatest challenge.[23] That might have been the case in previous elections, but he was proven wrong when the perfect storm of the recall hit.

48

5

Qualifying for the Ballot

Right-Wing Conspiracy or Convergence of Interests?

Thirty-one previous attempts to recall governors of California have failed. None even qualified for the ballot, although a 1968 effort to recall Governor Ronald Reagan came close. Yet in 2003, more than 1.3 million people readily signed valid petitions to put the recall of Governor Gray Davis before the voters. Why this time but not before? We've inventoried all the elements that created the "perfect storm," from the energy crisis, the recession, and the budget deficit to legislative gridlock and, finally, to the personality and leadership style of the governor.

But none of these elements was capable of going out to the malls and streets of California and collecting signatures. That took people—initially volunteers, soon thereafter paid workers. And someone had to get the whole thing started. As we noted earlier, California's multifaceted diversity ensures that for almost any cause there will be an instigator. Most such causes never make the ballot. That this one did indicates that more than a little band of zealots was involved. Some, including Gray Davis, alleged that the recall was "a vast right-wing conspiracy," echoing Hillary Clinton's complaint about what she saw as the persecution of President Bill Clinton. Or was it just the product of its citizen instigators, snowballing beyond their wildest expectations? Alternatively, was it the result of a unique convergence of conditions (the perfect storm) and the interests of a variety of individuals and organizations?

Qualifying for the Ballot

California's requirements for recalling statewide elected officials are the easiest among the eighteen states that permit recalls. Seven states require specific grounds for recall; California does not. Eleven of the eighteen states require a number of signatures equal to 25 percent or more of the votes cast

in the election of the official being recalled. California requires 12 percent (a number that was particularly low in 2003 because voter turnout was so low in 2002 when Gray Davis was unenthusiastically reelected). Most states allow 60 to 90 days for the circulation of petitions; California allows 160.

Even so, 897,158 signatures is a lot. The number rises to even more daunting levels because petition circulators must take into account invalid signatures from people who sign more than once, those who aren't registered to vote, mischievous signers, or just people whose handwriting is illegible. Generally, petition circulators aim to get 25 to 40 percent more signatures than are officially required to assure qualification for the ballot.

The process starts when proponents of a recall serve their target with a notice of their intentions, which they must also publish and file with the secretary of state. The notice of intention must include a statement of two hundred words or less on the reasons for the recall. Within a week of the filing of the notice, the official who is designated for recall may file a written response with the secretary of state, also limited to two hundred words. The secretary of state must then approve the recall petition forms, which follow a standard format. Each step is subject to strict deadlines (usually 7 to 10 days). Once all this is completed, recall supporters can officially begin collecting signatures. Both those who circulate and those who sign petitions must be registered voters.

Gray Davis was elected to a second term as governor on November 5, 2002. He was sworn in on January 6, 2003. Within weeks, opponents were plotting recall. On March 25, 2003, the secretary of state certified the recall petition for circulation. Advocates had 160 days to obtain 897,158 valid signatures to qualify for the ballot.

Instigators and Visionaries

Despite his recent election victory, Governor Davis's approval ratings in statewide polls had begun slipping disastrously as the magnitude of the state budget deficit became known. With the deficit estimated at $34.6 billion over 2 years, Davis proposed $21 billion in spending cuts and $8.3 billion in tax increases. His plan pleased no one at this early point in the budget process, but Democratic legislators resisted reductions in services and Republicans resisted tax increases. As the governor and legislators negotiated, however, some antitax activists and Republicans smelled blood.

Melanie Morgan mentioned recalling Davis on her radio talk show on December 30, 2002. Sal Russo, a Sacramento-based political consultant who had been an adviser to Bill Simon in his losing campaign against Davis, set up a recall website 1 month later, perhaps as much to harass the governor as

in anticipation of what would follow. Assemblyman Howard Kaloogian, a Christian-right Republican from San Diego County, also took up the cause.[1]

And as early as November 2002, immediately after Davis's reelection, antitax activist Ted Costa had begun talking seriously about recall with friends and allies such as conservative activist John Stoos and Bakersfield political consultant Mark Abernathy. According to Abernathy, they looked at the form a recall election would take, with all contenders on the ballot and no runoff election, and saw it as "a thing of beauty for Arnold."[2] In early February 2003, Costa announced a recall campaign on talk radio in Sacramento, urging listeners to drop by the offices of his organization, the People's Advocate, to sign the required initial notice of recall. Abernathy says "the parking lot outside Costa's office was full in thirty minutes." Over four hundred people showed up and others drove by honking their horns.[3] On February 5, Costa announced the campaign outside the state capitol, charging Davis with "gross mismanagement" of the energy crisis and the budget deficit.[4] On the same day, Assemblyman Kaloogian announced that he was forming the Recall Gray Davis Committee (www.recallgraydavis.com). It was Costa, however, who submitted the notice of recall with the signatures of 95 voters (65 are required) as proponents, and on March 25, Secretary of State Kevin Shelley certified the petition and formal signature gathering started. The advocates would have 160 days—until September 2—to collect 867,158 valid signatures and qualify the recall for the ballot. Figure 5.1 features the proponents' argument in favor of recalling Governor Davis and his response as they were presented on the recall petitions.

Ted Costa is an unassuming activist who is chief executive officer for the People's Advocate, an antitax organization founded by Paul Gann. A farm boy from the Sacramento area, the 62-year-old Costa still lives near Sacramento in suburban Citrus Heights, where he and his wife enjoy "gardening, gourding and taking care of their many animals."[5] He runs the People's Advocate organization (with a staff of three) from an office in a Sacramento strip mall behind a Krispy Kreme donut shop. A dedicated activist—as CEO he's paid just $38,000 a year—Costa has been involved in many initiative campaigns, from term limits to redistricting and protection of Proposition 13, the 1978 measure that launched a tax revolt that spread around the world.

Indeed, the roots of the People's Advocate and perhaps the recall run back to that earlier shaking of California's political firmament. The People's Advocate was founded in 1974 by the late Paul Gann to mobilize citizens against high taxes using the tools of direct democracy. In 1978, Gann allied with Howard Jarvis, another antitax activist, to campaign for property tax reform by initiative. Rapidly rising property tax bills had been an issue in California for nearly a decade. Double-digit inflation was driving up property values

Figure 5.1 **Proponent's Grounds for Recall and the Governor's Response**

Proponent's Grounds for Recall and the Governor's Response

The recall petition circulated throughout the state included arguments for and against a possible recall. For your information, we are providing an identical copy of these arguments below.

Proponent's Statement of Reasons

TO THE HONORABLE **GRAY DAVIS:** Pursuant to Section 11020, California Election Code, the undersigned registered qualified voters of the State of California, hereby give notice that we are the proponents of a recall petition and that we intend to seek your recall and removal from the office of Governor of the State of California, and to demand election of a successor in that office. The grounds for the recall are as follows: **Gross mismanagement of California Finances by overspending taxpayers' money, threatening public safety by cutting funds to local governments, failing to account for the exorbitant cost of the energy fiasco, and failing in general to deal with the state's major problems until they get to the crisis stage. California should not have to be known as the state with poor schools, traffic jams, outrageous utility bills, and huge debts....all caused by gross mismanagement.**

Governor's Answer to the Statement

<u>IF YOU SIGN THIS PETITION, IT MAY LEAD TO A SPECIAL ELECTION THIS SUMMER COSTING US TAXPAYERS AN ADDITIONAL $20-40 MILLION.</u>

Last November, almost 8,000,000 Californians went to the polls. They voted to elect Governor Davis to another term.

Just days after the Governor's inauguration in January, however, a handful of rightwing politicians are attempting to overturn the voters' decision. They couldn't beat him fair and square, so now they're trying another trick to remove him from office.

This effort is being led by the former Chairman of the State Republican Party, who was censured by his own party.

We should not waste scarce taxpayers' dollars on sour grapes. The time for partisanship and campaigning is past. It's time for both parties to work together on our State's problems.

Moreover, the allegations leveled against the Governor are false. As Governor, Davis has vetoed almost $9 BILLION in spending. California, along with 37 other states, is facing a budget deficit due to the bad national economy. The Bush Administration has announced the federal deficit this year will be the biggest in history, $304 BILLION.

In these difficult and dangerous times, LET'S WORK TOGETHER, not be diverted by partisan mischief.

Source: California Secretary of State, www.ss.ca.gov/elections/recall_notice.pdf.

and hence tax bills, while local governments, the primary recipients of property taxes, declined to cut tax rates proportionately as they struggled to meet growing service demands. A national study found that in 1942 California property taxes had been 89 percent of the national average, but by 1976 they had risen to 142 percent.[6] Always the most hated of taxes, the property tax became the prime target of antitax crusaders. Anger about property tax bills was exacerbated by news of a projected surplus in state funds. "Why did government need so much money, anyway?" people said, unaware of the different "pots" for funds at the state and local levels.

Escalating and unpredictable property tax bills were an irritant and sometimes a burden to many taxpayers, but they were a particular hardship for senior citizens on pensions or social security, whose incomes did not rise with inflation. Senior citizen organizations, antitax conservatives, and liberals concerned about the regressivity of the property tax all talked about reforms, albeit from very different perspectives. Two ballot measures were attempted, including one sponsored by then-Governor Ronald Reagan, but both failed. So while Sacramento debated, Paul Gann and Howard Jarvis acted.

Jarvis was the more visible of the two. A cantankerous senior citizen, Jarvis had run for office unsuccessfully several times on a vaguely populist but clearly antitax platform. A paid director of the Apartment Association of Los Angeles—condemned by critics as a representative of landlords—Jarvis caught the mood of the moment with his flamboyant behavior, including frequent use of the line from the 1976 movie *Network*, "I'm mad as hell and I'm not going to take this any more!"

The ebullient Jarvis generally overshadowed the more reserved Paul Gann. A native of Arkansas who came to California in the 1930s, Gann retired from selling cars and real estate to become an antitax crusader and a persistent government gadfly in the 1970s. Declaring himself "the people's advocate," he led the organization of the same name until his death in 1989. Membership was mostly based on and connected to the organization by direct mail but was nevertheless sufficiently formidable to qualify a number of significant initiatives for the California ballot.

But it was Proposition 13 that put Paul Gann, Howard Jarvis, and their organizations on the political map. They collected 1.2 million signatures to qualify for the ballot and—despite strong opposition by the state's political and business establishments—Proposition 13 passed by 65 to 35 percent. The measure immediately cut property taxes by $7 billion and set a strict cap on annual increases. The cuts hit cities, counties, and schools hardest and the state's role in local finance was greatly increased. Proposition 13 created as many problems as it solved, but antitax crusaders cheered and many ordinary citizens were happy. The tax revolt of 1978 shook

California's political firmament, and it soon spread to other states and even other nations.

Gann and the People's Advocate followed in 1979 with Proposition 4, which limited government spending increases to inflation and population increase and required that surplus taxes (if any) be returned to the people. Proposition 4 passed by 74 percent. In 1982, the People's Advocate co-sponsored another successful initiative called the Victim's Bill of Rights, toughening criminal sentencing. Ted Costa joined the organization that same year, working as Paul Gann's assistant until Gann's death in 1989, when Costa took over as leader. Under their leadership, the group campaigned for English as the official state language, legislative term limits, and other limits on taxation.

With a staff of three, the People's Advocate is not exactly a powerhouse in California politics. The Howard Jarvis Taxpayers Association, named for the leader who died in 1986, has had a higher profile under the leadership of Jon Coupal. The Jarvis group contributed to the recall by constant comment and criticism of the state's fiscal policy, particularly the controversial increase in license fees for motor vehicles.

While Costa is a hero to many, he also has his detractors. Bob Mulholland, an adviser to the California Democratic Party, labeled Costa "destructive" and "a kook." "He is living off the system of being anti-anti," Mulholland charged, exploiting senior citizens for small contributions. "He is not the Mother Teresa of citizen groups. This is a blood-sucking institution that he runs."[7] Consumer advocate Lenny Goldberg calls Costa "a right-wing nut-case."[8]

But it was Ted Costa and the People's Advocate that led the grassroots effort to qualify the recall for the ballot. Initially, they were pretty much out there on their own. "We wanted to have a little tea party here at the People's Advocate," Costa said. "We didn't have any tea, so we decided we'd throw the governor overboard."[9] A few individuals like Republican consultant Sal Russo and Assemblyman Howard Kaloogian were also advocating recall. Kaloogian reports, however, that Republican colleagues in the legislature "mocked" him and one, referring to his talk radio activities, called him "radioactive."[10] But while the Republican political establishment initially kept its distance from the recall, on February 23, delegates to the state Republican Party convention voted to support the effort.

Still, to most political observers, Costa's recall campaign looked like just another in a long line of unsuccessful efforts to recall governors. Gray Davis was in deeper trouble than any of his predecessors, but collecting the needed signatures to qualify a recall seemed an impossibly daunting task, particularly given the weak organizational infrastructure of its proponents and the apparent reluctance of Republican leaders to support it.

The Stealth Campaign: The Internet and Talk Radio

In the early months, recall efforts passed beneath the radar of mainstream politics, while two media outlets that have rarely had a significant impact on politics fueled it: the Internet and talk radio. In a huge state where communities are more virtual than real, these two communication networks brought together the disaffected in ways that over-the-fence backyard conversations might elsewhere.

Costa's group (www.davisrecall.com) and the Russo/Kaloogian group (www.recallgraydavis.com) had both established websites early in 2003. More than ever before, these websites reached voters and activists. Both sites were deluged with contacts from people curious about the recall; many downloaded petitions—a new way to distribute the forms and collect signatures. Costa claims that a million people downloaded petitions.[11] Russo reported 450,000 downloads of the petition and 20 million hits to their website.[12] While it is impossible to verify these numbers and the secretary of state did not distinguish downloaded petitions from those circulated in the traditional manner, it's clear that the Internet had a huge impact on the recall process, both in spreading the word and, perhaps more significantly, in collecting signatures. This could become a major tactic in direct democracy in the future and, if Howard Dean's use of the Internet in his campaign for the 2004 Democratic presidential nomination is any indication, in candidate campaigns as well. It is by far the fastest and cheapest way to reach supporters. In the past, the Internet was generally a passive campaign tool, merely making information available to people interested enough to find a campaign website. But the recall and Howard Dean campaigns prove that the Internet can motivate people to take action, not just seek information. E-mail lists have joined direct mail and donor lists as a valued campaign resource.

"We definitely hit a nerve," Sal Russo told the *California Journal*, "but you don't get that unless people are angry."[13] Someone or something was needed to stir that anger, to let people know about the recall option and get them to act on their anger. That's where talk radio came in.

Melanie Morgan and Michael Savage in San Francisco, Mark Williams and Eric Hogue in Sacramento, and Roger Hedgecock in San Diego were among the instigators and chief agitators for recall. Most of California's forty to fifty talk shows as well as nationally syndicated shows and their audiences joined in. Huge numbers of people listen to these programs, especially men. Many talk jockeys are skeptical about politics, some to the point of acting as gadflies. Most are conservative, but they are ideological conservatives rather than Republican Party loyalists. They thrive on anger and reaction. Liberal talk shows can be found on public radio, but they've never found the mass audience of their conservative counterparts whose explicit mission is to stir up their audiences.

In May, the infamous John and Ken in Los Angeles joined the right-wing talk shows in their obsession with recall. Broadcasting weekdays from 3 to 7 p.m., John Kobylt and Ken Chiampou reach most of Southern California during drivetime. Their politics are not purely conservative. They are as outraged by "crackpots" of the right (they cite Oliver North, Michael Reagan, and Gordon Liddy) as crackpots of the left, describing themselves as "rabid dogs." They latched onto the recall later than others, but, in their own words, "then the car tax happened: and the thing exploded."[14] With his approval ratings at an all-time low, Gray Davis was an easy target. But with real organizations promoting recall, the talk jockeys pushed their audiences beyond reaction to action. Download, sign, and circulate petitions!

All this was happening largely beyond the antennae of mainstream political leaders and journalists, few of whom listen to what they view as right-wing talk radio. When they did notice, Melanie Morgan says, "Almost from the beginning every liberal media outlet in the state referred to us as right-wing fanatics, a fringe political movement. We were derided as complete nutballs."[15]

But the "nutballs" in combination with the activists and the Internet were building a movement. "Historians will look back on this election as the time that the combination of the Internet and talk radio overtook the power of daily newspapers," says Shawn Steel, former chair of the state Republican Party and a leader of the recall movement.[16] Meanwhile, volunteers circulated petitions outside shopping malls, big-box retail outlets, and gun shops.

Yet for all the fury and all the enthusiasm, the collection of signatures did not initially produce the numbers that were necessary to put the recall on the ballot. In April, about two thousand signatures a day were coming in—a rate that meant the deadline could not be met. Republican leaders from the state legislature to the White House kept their distance, denying the movement validation while watching with interest. Successful statewide signature gathering usually costs millions of dollars, but Costa and the People's Advocate started with just $200,000. "That's a drop in the bucket for a major statewide campaign," says a Costa ally.[17] Small contributions were flowing in, but not enough for the major effort needed to meet the deadline.

Then in May, the final component was added. Ted Costa gives due credit to activists and talk radio, but he gives even more credit to the Internet and "the goose that laid the golden egg"[18]: Republican Congressman Darrell Issa, who provided the major funding the signature campaign needed.

Darrell Issa's Golden Egg

In early May, millionaire Republican Congressman Darrell Issa set up the third recall organization, Rescue California (www.rescuecalifornia.com), managed by Sacramento political consultant David Gilliard. At that point, when the other recall groups had collected about 100,000 signatures, California's politicians and political observers suddenly realized that a recall election could really happen. Issa announced that he would provide money for professional signature gatherers for the recall, starting with a check for $100,000. If Issa was willing to spend more than that—a lot more—political observers knew that the recall, like many esoteric initiatives, could actually make the ballot. Issa ultimately contributed nearly $2 million—two-thirds of the total spent by the three recall organizations. He also announced that he would be a candidate to replace Gray Davis as governor.

Issa made millions with his car alarm business, then used his wealth to get into politics. In 1998, he spent $10 million of his own money to seek the Republican nomination for U.S. Senate—unsuccessfully. Two years later, he won a seat in the U.S. House of Representatives in a conservative San Diego County district. A staunch conservative himself, Issa opposes abortion and gun control and helped fund the passage of Proposition 209 in 2000, the initiative banning affirmative action in California. Some saw Issa's involvement in the recall effort as opportunism; others wondered if Republican leaders in the House of Representatives or even the White House had put him up to it. Whatever his motives, his money changed everything. Political consultants and insiders had been watching the slow progress of the recall petitioners with skepticism, noting that a budget of around $2 million is usually necessary for an initiative to qualify for the ballot.

Without the money, many were doubtful that the recall would ever make the ballot. Talk radio, the Internet, and the persistence of activists like Ted Costa might have proven them wrong, but the date of the election and its outcome might have been different. By accelerating signature gathering, Issa guaranteed that the recall would be decided in a special election rather than being postponed to the upcoming March 2004 primary. With an expected hot contest for the Democratic presidential nomination on that ballot, a higher voter turnout among Democrats—presumed to be sympathetic to Governor Davis—could be expected. But that door was soon slammed shut.

As the recall drive heated up, volunteers were bringing in 5,000 signatures a day. Now, with Issa's money, paid professional signature gatherers were collecting 15,000 signatures a day.[19] The recall drive was in business.

Davis Fights Back

Although Governor Davis and his campaign strategists knew his anemic 2002 victory made him vulnerable, they seem to have been complacent up to this point, relying on traditional wisdom and expectations about underfunded petition drives. But when Issa started spending money, the Davis organization went into action.

On May 28, the formation of Taxpayers Against the Governor's Recall was announced. "The people behind the campaign to recall Governor Davis wear their partisan motives right out on their sleeves," said Dan Terry, president of the California Professional Firefighters Union and a leader of the anti-recall group. "An election was held. They lost, and now they want a 'do-over.'"[20] Labor unions, environmentalists, women's groups, and others rallied around Davis, and leaders of the Democratic Party joined in with varying degrees of enthusiasm (reflecting Davis's lack of friends and close allies). The group expected to raise $4 million to stop the recall.

Davis and his allies started by hiring the three top signature-gathering firms in California to circulate petitions supporting him. The intent was not to put anything on the ballot but rather to prevent these professional signature gatherers from going to work for Issa's Rescue California. The recall proponents found a professional to help them, however, by bringing Tom Bader, a former Californian and former operator of a petition business, back to the state.

By the end of June, signature gatherers for and against the recall were in direct competition. Opponents of the recall actively encouraged their people to disrupt signature gathering by the proponents by getting into conversations with them so they'd be diverted from potential signers. Opponents were also directed to urge citizens to complain to store managers about harassment by the pro-recall workers. Each side alleged that the other was misleading citizens about what they were signing. Insults, arguments, and fights were reported as the recall deteriorated, in some places, into hand-to-hand combat.[21]

Meanwhile, Davis marshaled his advisers. Garry South, his longtime political strategist, took the lead. Davis also brought in Chris Lehane and Peter Ragone, who had worked on national Democratic campaigns and who were in Florida for the battle of the chads that followed the 2000 presidential election. In the minds of many Democratic activists, the recall was an extension of exactly that kind of hardball Republican politics. Mainstream Democrats, led by U.S. Senator Dianne Feinstein, House Minority Leader Nancy Pelosi, and others, perhaps somewhat belatedly, formed a united front for Davis, along with labor and other groups. Business groups, including the powerful California

Business Roundtable, joined them. "When I talk to business leaders, to financiers, they say this is not about supporting the governor," said Art Pulaski, president of the California Federation of Labor. "This is about the damage the recall does to the state's effort to improve the economy and balance the budget."[22] In an opinion piece in the *Los Angeles Times,* labor and business leaders even pleaded with President Bush to oppose the recall campaign.

The Davis supporters also struck out at Darrell Issa. In early July, newspapers around the state ran stories reporting that in the 1970s Issa had been arrested on illegal-weapon charges twice. One of these arrests resulted in a misdemeanor conviction. Another story about a charge of auto theft also came out. Issa admitted the incidents but dismissed them as old news and typical Davis smear politics. Meanwhile, in the heat of questioning from journalists, he frequently gave confusing and inconsistent answers about his record on issues from gun control to abortion.[23]

Davis also attempted to label the recall activity part of "an effort by the right wing to overthrow the legitimate results of an election last November."[24] Later, he said "What's happening here is part of an ongoing national effort to steal elections Republicans cannot win."[25]

Republicans Engaged

The actions of Davis and his supporters may have backfired to the extent that they helped provoke the direct involvement of California's Republican leadership in the recall effort. Initially, party leaders had kept their distance from the recall, perhaps skeptical that it would succeed or wishing to avoid making it a purely partisan battle. But as the Davis defenders became more active and the recall signatures piled up, the Republican leadership joined in.

At the end of May, the first major Republican Party organization, the Lincoln Club of Orange County, endorsed the recall and contributed $50,000 to the campaign. By early June, Republican legislators were actively supporting recall. In July, Duf Sundheim, the chairman of the California Republican Party, announced the party's endorsement of the recall. Before committing, the party had conducted focus groups to assess public attitudes. "The story from these people is a fundamental thing," Sundheim said. "People are fed up with what's going on in California. It's unlike anything I've seen in California since Proposition 13."[26]

Sundheim made his statement from Washington, D.C., where he was meeting with President Bush's reelection team, leading to speculation that the Bush White House and particularly Karl Rove, the president's chief political operative, were behind the recall. The president said he thought Schwarzenegger, who had campaigned for his father, would make a good governor at some point. But with other Republicans likely to run, he said no more—at least in public. After the election, however, White House advisers admitted that President Bush had favored Schwarzenegger.[27] The Republican president may have been enjoying the spectacle of a Democratic governor on the ropes, but no evidence emerged of explicit White House involvement and Bush avoided discussion of the topic. National Republican operatives, like Frank Luntz, a pollster who helped Newt Gingrich develop his successful "Contract with America," were involved in the California recall effort, but so were national Democratic operatives. One skeptical Republican consultant dismissed charges of White House involvement on the grounds that "everything the White House does in California turns to s—," according to the *California Journal*.[28]

Meanwhile, media and voter interest in the recall were growing more intense. Movie star Arnold Schwarzenegger told *Esquire* magazine (whose cover he graced in July) that he'd love to be governor of California. "If the state needs me," he said, "and if there's no one I think is better, then I will run."[29] Speculation about other candidates, including leading Democrats, was rife.

An Election Is Ordered

On July 23, 2003, history was made. Through the hot summer day, people waited for an announcement from Secretary of State Kevin Shelley. Recall supporters marched and chanted outside the capitol building, demanding that Democrat Shelley certify the petitions but fearing he would find a way to delay the election to help fellow Democrat Gray Davis. Satellite vans for television stations around the state and nation were set up, ready to broadcast the news "*live* from the state Capitol." Nearby, two politically conservative talk radio hosts who had been instrumental in arousing the electorate to demand a recall set up shop and were already broadcasting. The atmosphere was a combination of a campaign rally, a circus, and a political lynching.

Then, at 5:30 p.m.—in time to go live on evening news programs—Secretary of State Kevin Shelley announced that the supporters of the recall had submitted more than enough valid signatures to qualify for an election—1.3 million of 1.7 million submitted (897,158 were required). This overwhelming number was indicative of the depth of discontent among the voters, not only with Gray Davis but also with legislative gridlock and with the general condition of politics and the economy.

But the discontent ran deeper in some parts of the state than others. As Figure 5.2 demonstrates, support for the recall was far stronger in some parts of California than others. San Bernardino and Riverside Counties alone provided more than 12 percent of the signatures. San Diego, Los Angeles, and

Figure 5.2 **Registered Voters Who Signed Recall Petitions**

Percent signing recall petitions

COUNTY	% OF VOTERS SIGNING				
Placer	20.5	Orange	14.7	Colusa	8.1
Madera	18.2	Fresno	13.7	Plumas	7.4
El Dorado	17.9	Mariposa	13.1	Yolo	7.3
Nevada	16.1	Tulare	12.9	Los Angeles	7.1
Sutter	16.1	San Diego	12.8	Sierra	7.0
Kern	16.0	Lassen	11.7	Mono	6.3
Shasta	15.4	Amador	11.5	Solano	6.3
San Bernardino	15.3	San Luis Obispo	11.3	Modoc	5.9
Tehama	14.8	Riverside	11.2	Alpine	5.8
		Kings	11.0	Santa Clara	5.6
Placer:		Glenn	10.6	Merced	5.1
1 out of 5		Sacramento	10.6	Napa	4.8
voters		Stanislaus	10.3	Contra Costa	4.7
signed		Yuba	10.1	Lake	4.7
petition		Tuolumne	9.8	Santa Barbara	4.4
		Siskiyou	9.6	Santa Cruz	4.4
		Ventura	9.5	Trinity	4.4
		Calaveras	9.0	Sonoma	4.3
		Butte	8.9	Imperial	4.1
		San Joaquin	8.6	San Mateo	4.0
		Inyo	8.5	Marin	2.9
		Humboldt	8.4	Del Norte	2.8
				San Benito	2.6
				Alameda	2.1
				Mendocino	2.0
				San Francisco	1.0
				Monterey	0.0

Percent of registered voters in each county who signed petitions in support of the recall of Gov. Gray Davis:

- ☐ 0-5%
- ☐ 5.1-9%
- ■ 9.1-14%
- ■ 14.1-21%

MERCURY NEWS

Orange County were also major contributors. The highest percentage of registered voters signing was in some of the Gold Country and mountain counties with small populations and a conservative orientation. More important, roughly 15 percent of registered voters signed petitions in San Bernardino and Orange Counties, followed by Fresno (13.7 percent), San Diego (12.8 percent), Riverside (11.2 percent), and Sacramento (10.6 percent). Support for the recall was weaker in Los Angeles (7.1 percent) and Santa Clara Counties (5.6 percent) and weaker still in San Francisco (1 percent) and other Bay

Area counties.[30] Predictably, the more conservative counties supported the recall with the greatest enthusiasm. Perhaps less predictably, these numbers prefigured the actual election outcome.

The effort to qualify the recall for the ballot was only the beginning, of course. An intense campaign of just 75 days would follow. But none of this would have happened without the efforts of thousands of people tapping into a profound discontent among the voters. Two individuals, however, were essential. Darrell Issa injected the funds that accelerated the process. Bruce Cain, director of the Institute of Governmental Studies at the University of California, Berkeley, says "Issa's lasting contribution to California politics is that he proved that the recall behaves the same way as regular initiatives in that you can get them before voters if you spend enough money."[31] But the most essential individual, clearly, was Ted Costa, the modest activist who started it all. "People like Ted light matches," says Republican consultant and recall advocate Sal Russo, "but other factors may determine if those matches ignite anything."[32] Clearly those factors were present when Costa struck his match.

Part III

The Campaign

6

The 75-Day Sprint

The typical election cycle for a statewide election in California is 168 days. That's the time when candidates take out and complete the filing of papers with the secretary of state, raise and spend the bulk of their campaign funds, and present their cases to the voters. During the same period, the secretary of state prepares the various ballot descriptions and county registrars organize their precinct operations. Most people don't pay attention to these minutiae, but it's a busy time for all involved. So, imagine how state and local officials must have felt when they learned that they would have less than half the usual time to prepare for the recall election. They were not happy campers; they also had no choice.

According to California's recall law, once the secretary of state certifies the required number of signatures as valid, the lieutenant governor is required to call an election within 60 to 80 days. That's precisely what Democratic Lieutenant Governor Cruz Bustamante did on July 25, 2003, when he scheduled the election for October 7, 2003, or 75 days after the certification declared by Democratic Secretary of State Kevin Shelley. With Bustamante's proclamation, the recall campaign was officially under way and Gray Davis was officially on notice.

But there were other problems for Davis. Whereas the federal impeachment process requires the accused to be confronted with criminal charges, California, like most recall states, attaches no particular conditions to the process. In other words, basic discontent with the incumbent is enough to lead to a recall effort—and there was plenty of that for the embattled Davis.

And so the stage was set for what was sure to be frenetic activity. After Bustamante set the October 7 date, the political dam burst. Within days, 472 would-be candidates took out filing papers which, according to state law, were due by August 9, a mere 17 days after the announcement. The requirements for an individual to be put on the replacement ballot were simple enough—$3,500 and the signatures of 65 registered voters. By the closing date, 247 people had filed candidacy papers; of that number, the secretary of state declared that 135 individuals fulfilled the filing conditions. The great race was on.

The Recall Goes to Court

Like so many issues in California politics, the recall effort soon wound up in the courts. First recall advocates sued Secretary of State Kevin Shelley (a Democrat) on grounds that he was slowing down the validation of signatures so that the election would be held in March, at the same time as the Democratic presidential primary. Then recall opponents filed a suit demanding a judicial order to stop the process until the credentials of the signature gatherers as registered California voters (as required by law) could be verified. The courts rejected the latter case and, in the former, directed the state to order counties to verify signatures as they were filed rather than waiting for them to accumulate. Both rulings favored the recall proponents. On July 24, the California State Supreme Court dismissed all arguments against the recall qualifying for the ballot.

Additional cases nevertheless followed in the federal courts. Each one caused voters and observers to gasp and worry (or hope) that the election would be halted or delayed. A threat to do so arose in Monterey County, where the federal Voting Rights Act requires Justice Department approval of elections and where it was alleged that the rights of minority voters would be limited by a plan to consolidate several voting places into one for economic reasons. Such polling places would be less accessible to voters than the traditional neighborhood voting sites. A judge warned that the election could be delayed if Justice Department approval was not forthcoming—but it was and the case went away.

The American Civil Liberties Union (ACLU) filed a case in August arguing that as many as 40,000 people would be disenfranchised by punch-card voting machines and faulty chads (shades of Florida!). The case was dismissed in the U.S. District Court but appealed to the Ninth Circuit Court of Appeals. On September 15, a three-judge panel of the Court of Appeals overturned the lower court ruling and declared that the use of punch-card machines in six counties with 44 percent of the state's voters would violate the "equal protection" clause of the U.S. Constitution because it would disproportionately affect minority voters. The election was rescheduled for March 2, pending a possible appeal to the U.S. Supreme Court. Instead, a sufficient number of members of the Ninth Circuit Court of Appeals called for an en banc hearing on the case, with a panel consisting of the Chief Justice of the Ninth Circuit and ten associates selected at random. Oral arguments were heard on September 22—just over 2 weeks before the scheduled election—and on September 23, the Ninth Circuit Court of Appeals overruled the three-judge panel in a stunning 11-to-0 vote, restoring the election date of October 7.

Even though the pro-recall activists and voters won, the court cases probably added to their anger, as many of them were still upset about previous court rulings overturning initiatives (state and federal courts have frequently found California initiatives to be unconstitutional). The court cases also added to the uncertainty and drama of the recall election. Because it was the first statewide recall in California history, much was unknown about the process, both legally and politically. The court rulings resolved much of the uncertainty, but not without much trepidation on the part of both proponents and opponents of the recall.

A Two-Part Affair

Far from a traditional election, the recall presented rules and processes different from anything Californians had ever known. To begin with, state law provides for the recall and replacement election simultaneously. The idea is that if the voters actually recall the incumbent governor, a successor should be ready to step in. Thus, the ballot has two questions. The first part reads, "Shall (name) be recalled from the office of governor," with voter options of "yes" and "no." The second part simply lists the names of the candidates who have been certified as eligible by the secretary of state. At first, statewide surveys showed confusion among sizable numbers of voters over the two-part ballot, with many thinking that they could participate in only one part of the ballot. By the time of the election, however, voters were clear on the process. Moreover, according to a statewide poll, 90 percent of the state's voters were following the race closely.[1]

The two-part ballot presented an even more potentially serious problem for Davis, however. Unlike the law of most recall states, California's does not allow the recalled governor to run on the ballot with those who would succeed him. In other words, if the "yes" votes prevailed on the first part of the recall ballot, Davis would be out. As to the second part, recall law provides that whoever gets more votes than anyone else wins the election, no matter how small or large the number of votes the winner receives. All of this presented a curious possibility that, should Davis be recalled from office, his successor might actually gather fewer votes in a crowded race of replacement candidates than the recalled governor himself. And crowded it was, although by October 7, the list of viable candidates was pared down to two.

Gray Davis: Defending Slippery Turf

If Gray Davis could attract a majority of "no" votes on the first part of the recall ballot, then the second part and its attendant humiliation would not

Figure 6.1 **Official Ballot**

Statewide Special Election
Orange County, California
October 07, 2003

OFFICIAL BALLOT

SAMPLE BALLOT

Instruction Note:

HOW TO VOTE:
- To vote, fill in and BLACKEN completely the rectangle to the left of any candidate or to the left of the word "YES" or "NO".
- Vote for only ONE of the 135 candidates, OR enter a write-in candidate in the space provided.
- Use only the special marking device provided. (Absentee voters should use a dark pen or a #2 pencil.)

Shall GRAY DAVIS be recalled (removed) from the office of Governor?

☐ YES

☐ NO

Candidates to succeed GRAY DAVIS as Governor if he is recalled:
Vote for One

☐ B.E. SMITH
Independent - Lecturer

☐ DAVID RONALD SAMS
Republican - Businessman/Producer/Writer

☐ JAMIE ROSEMARY SAFFORD
Republican - Business Owner

☐ LAWRENCE STEVEN STRAUSS
Democratic - Lawyer/Businessperson/Student

☐ ARNOLD SCHWARZENEGGER
Republican - Actor/Businessman

☐ GEORGE B. SCHWARTZMAN
Independent - Businessman

☐ MIKE SCHMIER
Democratic - Attorney

☐ DARRIN H. SCHEIDLE
Democratic - Businessman/Entrepreneur

☐ BILL SIMON
Republican - Businessman

☐ RICHARD J. SIMMONS
Independent - Attorney/Businessperson

☐ CHRISTOPHER SPROUL
Democratic - Environmental Attorney

☐ RANDALL D. SPRAGUE
Republican - Discrimination Complaint Investigator

☐ TIM SYLVESTER
Democratic - Entrepreneur

☐ JACK LOYD GRISHAM
Independent - Musician/Laborer

☐ JAMES H. GREEN
Democratic - Firefighter Paramedic/Nurse

☐ GARRETT GRUENER
Democratic - High-Tech Entrepreneur

☐ GEROLD LEE GORMAN
Democratic - Engineer

☐ RICH GOSSE
Republican - Educator

☐ LEO GALLAGHER
Independent - Comedian

☐ JOE GUZZARDI
Democratic - Teacher/Journalist

☐ JON W. ZELLHOEFER
Republican - Energy Consultant/Entrepreneur

☐ PAUL NAVE
Democratic - Businessman/Prizefighter/Father

☐ ROBERT C. NEWMAN II
Republican - Psychologist/Farmer

☐ BRIAN TRACY
Independent - Businessman/Consultant

☐ A. LAVAR TAYLOR
Democratic - Tax Attorney

☐ WILLIAM TSANGARES
Republican - Businessperson

☐ PATRICIA G. TILLEY
Independent - Attorney

☐ DIANE BEALL TEMPLIN
American Independent - Attorney/Realtor/Businesswoman

☐ MARY "MARY CAREY" COOK
Independent - Adult Film Actress

☐ GARY COLEMAN
Independent - Actor

☐ TODD CARSON
Republican - Real Estate Developer

☐ PETER MIGUEL CAMEJO
Green - Financial Investment Advisor

☐ WILLIAM "BILL" S. CHAMBERS
Republican - Railroad Switchman/Brakeman

☐ MICHAEL CHELI
Independent - Businessman

☐ ROBERT CULLENBINE
Democratic - Retired Businessman

☐ D. (LOGAN DARROW) CLEMENTS
Republican - Businessman

☐ S. ISSA
Republican - Engineer

☐ BOB LYNN EDWARDS
Democratic - Businessman

☐ ERIC KOREVAAR
Democratic - Scientist/Businessman

☐ STEPHEN L. KNAPP
Republican - Engineer

☐ KELLY P. KIMBALL
Democratic - Business Executive

☐ D.E. KESSINGER
Democratic - Paralegal/Property Manager

☐ EDWARD "ED" KENNEDY
Democratic - Businessman/Educator

☐ TREK THUNDER KELLY
Independent - Business Executive/Artist

☐ JERRY KUNZMAN
Independent - Chief Executive Officer

☐ PETER V. UEBERROTH
Republican - Businessman/Olympics Advisor

☐ BILL PRADY
Democratic - Television Writer/Producer

☐ DARIN PRICE
Natural Law - University Chemistry Instructor

☐ GREGORY J. PAWLIK
Republican - Realtor/Businessman

☐ LEONARD PADILLA
Independent - Law School President

☐ RONALD JASON PALMIERI
Democratic - Gay Rights Attorney

☐ CHARLES "CHUCK" PINEDA JR.
Democratic - State Hearing Officer

☐ HEATHER PETERS
Republican - Mediator

☐ ROBERT "BUTCH" DOLE
Republican - Small Business Owner

☐ SCOTT DAVIS
Independent - Business Owner

☐ RONALD J. FRIEDMAN
Independent - Physician

☐ GENE FORTE
Republican - Executive Recruiter/Entrepreneur

☐ DIANA FOSS
Democratic -

☐ LORRAINE (ABNER ZURD) FONTANES
Democratic - Film Maker

☐ WARREN FARRELL
Democratic - Fathers' Issues Author

☐ DAN FEINSTEIN
Democratic -

☐ LARRY FLYNT
Democratic - Publisher

☐ CALVIN Y. LOUIE
Democratic - CPA

☐ DICK LANE
Democratic - Educator

☐ TODD RICHARD LEWIS
Independent - Businessman

☐ GARY LEONARD
Democratic - Photojournalist/Author

☐ DAVID LAUGHING HORSE ROBINSON
Democratic - Tribal Chairman

☐ NED ROSCOE
Libertarian - Cigarette Retailer

☐ DANIEL C. "DANNY" RAMIREZ
Democratic - Businessman/Entrepreneur/Father

☐ CHRISTOPHER HANKEN
Democratic - Planning Commissioner

☐ JEFF RAINFORTH
Independent - Marketing Coordinator

☐ KURT E. "TACHIKAZE" RIGHTMYER
Independent - Middleweight Sumo Wrestler

☐ DANIEL W. RICHARDS
Republican - Businessman

☐ KEVIN RICHTER
Republican - Information Technology Manager

☐ REVA RENEE RENZ
Republican - Small Business Owner

☐ SHARON RUSHFORD
Independent - Businesswomen

☐ GEORGY RUSSELL
Democratic - Software Engineer

☐ MICHAEL J. WOZNIAK
Democratic - Retired Police Officer

☐ DANIEL WATTS
Green - College Student

☐ NATHAN WHITECLOUD WALTON
Independent - Student

☐ MAURICE WALKER
Green - Real Estate Appraiser

☐ CHUCK WALKER
Republican - Business Intelligence Analyst

☐ LINGEL H. WINTERS
Democratic - Consumer Business Attorney

☐ C.T. WEBER
Peace and Freedom - Labor Official/Analyst

☐ JIM WEIR
Democratic - Community College Teacher

☐ BRYAN QUINN
Republican - Businessman

☐ MICHAEL JACKSON
Republican - Satellite Project Manager

☐ JOHN "JACK" MORTENSEN
Democratic - Contractor/Businessman

☐ DARRYL L. MOBLEY
Independent - Businessman/Entrepreneur

☐ JEFFREY L. MOCK
Republican - Business Owner

☐ BRUCE MARGOLIN
Democratic - Marijuana Legalization Attorney

☐ GINO MARTORANA
Republican - Restaurant Owner

☐ PAUL MARIANO
Democratic - Attorney

☐ ROBERT C. MANNHEIM
Democratic - Retired Businessperson

☐ FRANK A. MACALUSO, JR.
Democratic - Physician/Medical Doctor

☐ PAUL "CHIP" MAILANDER
Democratic - Golf Professional

☐ DENNIS DUGGAN MCMAHON
Republican - Banker

☐ MIKE MCNEILLY
Republican - Artist

☐ MIKE P. MCCARTHY
Independent - Used Car Dealer

☐ BOB MCCLAIN
Independent - Civil Engineer

☐ TOM MCCLINTOCK
Republican - State Senator

☐ JONATHAN MILLER
Democratic - Small Business Owner

☐ CARL A. MEHR
Republican - Businessman

☐ SCOTT A. MEDNICK
Democratic - Business Executive

☐ DORENE MUSILLI
Republican - Parent/Educator/Businesswoman

☐ VAN VO
Republican - Radio Producer/Businessman

☐ PAUL W. VANN
Republican - Financial Planner

☐ JAMES M. VANDEVENTER, JR.
Republican - Salesman/Businessman

☐ BILL VAUGHN
Democratic - Structural Engineer

☐ MARC VALDEZ
Democratic - Air Pollution Scientist

☐ MOHAMMAD ARIF
Independent - Businessman

☐ ANGELYNE
Independent - Entertainer

☐ DOUGLAS ANDERSON
Republican - Mortgage Broker

☐ IRIS ADAM
Natural Law - Business Analyst

☐ BROOKE ADAMS
Independent - Business Executive

☐ ALEX-ST. JAMES
Republican - Public Policy Strategist

☐ JIM HOFFMANN
Republican - Teacher

☐ KEN HAMIDI
Libertarian - State Tax Officer

☐ SARA ANN HANLON
Independent - Businesswoman

☐ IVAN A. HALL
Green - Custom Denture Manufacturer

☐ JOHN J. "JACK" HICKEY
Libertarian - Healthcare District Director

☐ RALPH A. HERNANDEZ
Democratic - District Attorney Inspector

☐ C. STEPHEN HENDERSON
Independent - Teacher

☐ ARIANNA HUFFINGTON
Independent - Author/Columnist/Mother

☐ ART BROWN
Democratic - Film Writer/Director

☐ JOEL BRITTON
Independent - Retired Meat Packer

☐ AUDIE BOCK
Democratic - Educator/Small Businesswoman

☐ VIK S. BAJWA
Democratic - Businessman/Father/Entrepreneur

☐ BADI BADIOZAMANI
Independent - Entrepreneur/Author/Executive

☐ VIP BHOLA
Republican - Attorney/Businessowner

☐ JOHN W. BEARD
Republican - Businessman

☐ ED BEYER
Republican - Chief Operations Officer

☐ JOHN CHRISTOPHER BURTON
Independent - Civil Rights Lawyer

☐ CRUZ M. BUSTAMANTE
Democratic - Lieutenant Governor

☐ CHERYL BLY-CHESTER
Republican - Businesswoman/Environmental Engineer

☐ Write-In

MEASURES SUBMITTED TO THE VOTERS

STATE

Proposition 53

FUNDS DEDICATED FOR STATE AND LOCAL INFRASTRUCTURE. LEGISLATIVE CONSTITUTIONAL AMENDMENT.

Generally dedicates up to 3% of General Fund revenues annually to fund state and local (excluding school and community college) infrastructure projects. Fiscal Impact: Dedication of General Fund revenues for state and local infrastructure. Potential transfers of $850 million in 2006-07, increasing to several billions of dollars in future years, under specified conditions.

☐ YES

☐ NO

Proposition 54

CLASSIFICATION BY RACE, ETHNICITY, COLOR, OR NATIONAL ORIGIN. INITIATIVE CONSTITUTIONAL AMENDMENT.

Prohibits state and local governments from classifying any person by race, ethnicity, color, or national origin. Various exemptions apply. Fiscal Impact: The measure would not result in a significant fiscal impact on state and local governments.

☐ YES

☐ NO

BT8

matter. That "if" became more illusory with each passing day of the recall campaign, for the Davis defense could never gain traction. There had been thirty-one attempts to recall California governors in the past, but this bad dream was quickly becoming a nightmare.

At first, Davis paid little public attention to the recall effort. Rather than respond to the fledgling signature-gathering effort, he attempted to portray himself as a captain trying to steer his ship through rough seas; nothing would keep him from finding safe harbor, and his passengers (the voters) would ultimately be grateful for his efforts. Nevertheless, the public associated Davis with bad news again and again. Periodically, he would announce new numbers on the deficit, with the new figure always larger than the previous one. He would then ask the legislature to move quickly on increasing selected taxes or cutting expenditures, yet the legislature moved little during the spring of 2003.

Nevertheless, for a while it appeared that Davis might beat back the recall effort. In April, 6 weeks into the signature-gathering effort, the recall seemed to stall. Ted Costa, one of the original leaders of the movement, fretted that the campaign wouldn't have enough financial support to pay people to gather the necessary signatures. Davis supporters crowed. "Nobody's wants to put any resources behind this thing [the recall] because they don't think it's the smart thing to do,"[2] gloated Davis spokesperson Roger Salazar. That gloat soon turned glum when Republican Congressman Darrell Issa started pouring funds into the recall effort. Eventually, Issa dumped nearly $2 million of his own money into the campaign, a critical element for recall forces desperately in search of funds. Now, in addition to large numbers of angry people, the recall forces had the necessary funds to hire paid signature gatherers, and the political environment changed considerably.[3]

Meanwhile, Davis continued to view the recall effort as a problem that could be overcome. He had good reason to take such a position. While a Field Poll released in mid-May showed his popularity at only 24 percent, nearly 60 percent of the respondents viewed the recall concept as a "bad idea." Counting on the fairness issue as a cornerstone of his anti-recall campaign, Davis hoped to ride out the political storm. But he also knew that the framing of election issues could play a major role in the outcome. If the election revolved around the policies of the Davis administration, his unpopularity would almost certainly lead to defeat. However, if the election were viewed as a conspiracy to undo the will of the voters from the previous November and undermine the democratic process, then he would have a chance. He pursued the fairness approach throughout July. But he had to be careful. Already possessing the reputation of a vicious campaigner, Davis had to make sure that he did not appear to be a bully. For the moment, he chose his words carefully.

Meanwhile, throughout May and early June, Davis reassembled his successful campaign team from 1998 and 2002 as he sharpened his anti-recall pitch. Speaking before the *Los Angeles Times* editorial board in late May, Davis accused Issa of financing a campaign to put himself into the governorship through the back door of the recall.[4] In late June, even before the measure qualified for the ballot, the governor described the recall effort as "partisan mischief by the right wing [of the Republican party]."[5]

At the heart of the Davis strategy was the question of culpability. How much was the governor responsible for the state's problems? More important, how much responsibility was he willing to own? With polls throughout the summer showing solid majorities supporting the recall effort, Davis answered these questions in late August. In a 19-minute speech televised statewide, he began by acknowledging that he could have done a better job of holding down expenses. But rather than make contrition a central theme, he quickly assigned blame for the state's difficulties to his predecessors (the energy crisis), the federal government (poor management of the weak economy), and a right-wing power grab (by the state's conservative sore losers).[6] The counterattack did not resonate with the voters. Instead of framing the issue on his terms, Davis made himself the issue. And he was running out of time to change the public's perception.

Shortly after the entry of Arnold Schwarzenegger as a replacement candidate, Davis shifted gears yet again. In some respects, the election cacophony became greatly reduced throughout August and early September; at the last minute, Darrell Issa decided not to file and Schwarzenegger swaggered in. Throughout the next few weeks, several top-tier candidates withdrew, although their names remained on the ballot. With Schwarzenegger now viewed as the major opponent, Davis struggled for a way to overcome the actor's sudden political stardom. One last chance emerged when, with 6 days remaining in the campaign, the *Los Angeles Times* published a series of startling stories that featured accusations from sixteen women—some going back as far as 25 years—that Schwarzenegger had groped them. Despite his reputation as a "no holds barred" campaigner, Davis held back, saying on October 2 that "the voters will determine how significant the [groping] issue is."[7] But the next day, Davis did an about-face. He demanded that Schwarzenegger apologize directly to his women accusers and insisted that the surging Republican candidate was unfit to govern.[8]

Davis's attack backfired. An NBC/Knight Ridder poll, taken on the Wednesday through Saturday before the election, showed great initial harm to Schwarzenegger as a result of the revelations, particularly among female voters. Female support for the recall drifted downward over the 4-day period, from 57 to 42 percent. But by deriding Schwarzenegger, Davis lost his brief advantage and Schwarzenegger gained a sympathy vote. By election day, women voters reversed course. Still worse for Davis, an NBC exit poll of voters found that of those deciding during the final week of the campaign—the week thought to mean bad news for Schwarzenegger—57 percent opted for the recall.

In the end, Schwarzenegger's candidacy may not have been as harmful to Davis as the governor was to himself. While the state plunged into economic despair, Davis seemed remote from the experience, even though his work ethic and personal values suggested the contrary. Writing 2 days before the end of the 75-day sprint, *Los Angeles Times* reporter David Zucchino seemed to summarize the governor's angst in noting that "Gray Davis is a man running against himself—against his image as aloof, cautious, condescending, and emotionally stunted."[9] All through the campaign Davis tried to project a more relaxed and human image and he did loosen up a bit, speaking more effectively to campaign-organized "town hall meetings"; but even before friendly crowds, Davis couldn't really escape his measured tones and stiff demeanor.

Arnold Schwarzenegger: Capitalizing on the Mystique of Celebrity

The entry of Arnold Schwarzenegger into the race to replace Davis was a dream come true for the media. Until the moment of his announcement, Schwarzenegger had told reporters—and even his own campaign consultants —that he would refrain from entering the contest in deference to former Los Angeles Mayor Richard Riordan, once the front runner in the 2002 Republican primary until Davis attack ads tilted the outcome in favor of conservative Bill Simon.

On August 6, only 3 days before the filing deadline, Schwarzenegger scheduled an appearance on the *Tonight Show*, ostensibly to explain why he wasn't going to run and why he would defer to long-time friend Riordan. In a vintage Hollywood moment detailing to host Jay Leno his thoughts about the recall, Schwarzenegger declared, "The politicians are fiddling, fumbling and failing, and the man who is failing more than anyone is Gray Davis. He's failing them terribly, and this is why he needs to be recalled. And this is why I'm going to run for governor of California."[10] The crowd burst into cheers and applause as Schwarzenegger orchestrated the first of many well-timed moves to victory. On August 9, the last day of the filing period, Schwarzenegger filed his papers before a national audience of millions of viewers.

Who is Arnold Schwarzenegger? The Davis camp tried to paint him as a one-dimensional action figure who was attempting to leverage his entertainment-generated popularity with the voters. Yet that was only part of the Schwarzenegger persona. Living what he described as the "immigrant's dream," the Austrian-born Schwarzenegger came to the United States as a teenager with $20 in his pocket, put himself through college, and began his now-famous bodybuilding career. From there he became a successful actor. But there were other sides to Schwarzenegger. He became involved in several business ventures and grew more active in politics. As early as 1984, he was active in national Republican campaigns. His 1986 marriage to Maria Shriver, a television journalist and member of the Kennedy clan, drew Schwarzenegger further into the political world. In 2002, he led the way on California's Proposition 49, an initiative that provided surplus state budget money for after-school programs. The initiative passed handily, although no surplus dollars have appeared. Whatever else Schwarzenegger was or was not, he had had exposure to politics in high places.

Indeed, the Schwarzenegger campaign resonated like no other. From his televised appearance on the *Tonight Show* until his election, Schwarzenegger leveraged his movie stardom to political success with masterful, well-planned strokes. Repeatedly, his well-known movie lines became the foundation of his political statements. During the Leno interview, for example, Schwarzenegger warned that elected officials needed to meet the public's expectations, "otherwise you are 'hasta la vista baby,'" referring to a line from his successful role in the *Terminator* movies.[11] As *San Francisco Chronicle* political reporter Carla Marinucci observed, his appearance on Leno's show "had all the elements of another hit movie: suspense, humor and drama, a great script—a surprise Hollywood ending."[12]

But it was more than just drama or the use of clever lines. Schwarzenegger's overall strategy underscored a fundamental point: He used his celebrity as political capital—in place of the traditional building blocks on which seasoned politicians depend. Whereas other major candidates relied on conventional endorsements, phone banking, and historic voting loyalties, Schwarzenegger counted on the personal connection between his movies and his fans. This populist approach gave star-struck voters the feeling that he was talking to them directly.

Despite the informality, there were few ad-libs in Schwarzenegger's campaign. Like a well-scripted movie, every aspect of the effort was carried out with discipline and precision. Early on, he committed to one political debate from the half dozen scheduled among the top-tier candidates, claiming that he would be too busy campaigning and "meeting with the voters." The debate in which he participated was the only one where the candidates knew the questions in advance. Opponents screamed that he was choreographing his campaign and not subjecting himself to the rigors of strict scrutiny; Schwarzenegger stayed on and within his own schedule and campaign design.

Accusations came and went. Nothing stuck, whether it was past statements, past behaviors, or past voting records. And the allegations seemed endless, ranging from illegal entry[13] to being a Nazi sympathizer[14] to having a poor voting record.[15] When he was accused of being a sexist, Schwarzenegger and his wife, NBC correspondent Maria Shriver, appeared on the *Oprah Winfrey Show*, where 86 percent of the viewers are female. With his wife next to him, Schwarzenegger explained his earlier public statements about women as playful, harmless attempts to be friendly.[16] Comedian David Letterman seemed to put the entire issue to rest when, reminding his national audience of former president Bill Clinton's dalliances, said, "I'm telling you, this guy [Schwarzenegger] is presidential material."[17] Quipped fellow comedian Jay Leno, "You've got Arnold, who groped a few women, or Davis, who screwed the whole state."[18] And just like that, the issue was neutralized.

But more than his defense was Schwarzenegger's ability to make his case before a national audience in a controlled environment. Even the groping accusations cited in the *Los Angeles Times* met with more—not less—voter support. Movie stardom had its benefits, about that there was little doubt.

As Schwarzenegger fought off attacks against his character, he defined himself as a political moderate in line with the values of most Californians. He kept his speeches brief and offered few details other than the big picture on major topics. People were less concerned about specifics, he would say, and more interested in leadership and "action, action, action."[19] And with that he straddled the line. On the question of abortion, for example, he identified with a woman's right to choose what to do about an unwanted pregnancy. When asked about gun control, he agreed that the state had the right to control the conditions of gun ownership. And on the question of gay rights, Schwarzenegger stated that he had no opposition to the issue. Each of these positions endeared Schwarzenegger to sizable elements of the moderate electorate. At the same time, he reached out to conservatives when he railed against more taxes, demanded reversal of the automobile license fee increases, and called for California to become business-friendly again. The fact that Schwarzenegger managed to hold onto both conservatives and moderates was a harbinger of the kind of broad support he would develop by the campaign's end. As conservative commentator Rush Limbaugh said, "Arnold Schwarzenegger is not a conservative. . . . That does not mean that he is not worthy."[20]

Meanwhile, Schwarzenegger had his own points to make about the Davis administration. Referring to the "pay to play" accusations against Gray Davis, he argued that "special interests" had too much power in Sacramento. In between signing autographs at the state fair in Sacramento, Schwarzenegger claimed that the state was overregulated and overtaxed. The problem? "It is all about politics. . . . What we have to do is get rid of an administration that is in the pockets of all the special interests."[21] Yet the candidate stayed away from specifics. When asked how he would change the direction of the state, Schwarzenegger responded, "What the people want to hear is: Are you willing to make changes? Are you tough enough to go in there and provide leadership? That's what this is about. I will be tough enough. And independent. I can go up there [to Sacramento] and really clean house."[22]

And clean house he did on October 7.

Cruz Bustamante: The Democratic Alternative

In many ways, he was the Cinderella candidate, although there would be no "happy ending" as in the fairy tale of the same name. Democratic Lieutenant Governor Cruz Bustamante enjoyed a unique position in the recall campaign. Whereas other key Democrats decided not to run as replacement candidates, Bustamante did choose to do so. By the time of the filing deadline, he stood out as the only well-known Democrat running for governor should the incumbent actually be turned out of office.

For some Democratic activists, Bustamante was an obstructionist to the process; after all, as a recall replacement candidate, he threatened any hopes of Democratic unity behind Democratic Governor Gray Davis. For others, Bustamante's candidacy represented the best hope of overcoming a fatally flawed Gray Davis, who would not survive the recall effort. Regardless of the interpretation, there was no doubt about one fact: Because of his unique "No on Recall, Yes on Bustamante" position, Bustamante himself became a source of considerable controversy. This strife would dog him throughout the campaign. But first the story . . .

When Bustamante announced the October 7 date of the recall election as constitutionally prescribed, he also stated that under no circumstances would he be a candidate. But within a week of his initial press conference, Bustamante did an about-face and announced that he *would* be a replacement candidate in the event that the governor was recalled. Why the sudden shift? Because, Bustamante claimed, polls showed the governor slipping, and he wanted to be the Democratic alternative should the recall succeed.

Bustamante found himself in a tough position. Unlike the national party nominations for president and vice president, California nominees for statewide offices are not linked. Thus, even though Bustamante was elected independently from the governor, he was a Democrat and still the number-two officeholder in the state. On the other hand, having been elected independently, Bustamante had the opportunity to distinguish himself from the governor. But could he pull it off? That would be his challenge.

74

That Bustamante was even in the race was a typical California miracle in itself. The Latino son of a barber and raised near Fresno, he spent his summers as a youth picking crops—something that millions of the state's Latinos had done themselves. Ironically, he seemed the perfect foil for the dapper Schwarzenegger. "I am short, I am overweight, and I am losing all my hair," he often said during the campaign, as if to cast himself as the antipolitician. He was anything but. Having worked for a congressman and then an assemblyman, Bustamante was elected to the assembly, then became Speaker, and ultimately rose to lieutenant governor in 1998, making him the first statewide-elected Latino in 125 years. Perhaps this would be his moment, planned or otherwise.

After declaring his candidacy, Bustamante quickly positioned himself as the liberal alternative to the status quo. He came out in favor of a bill to allow drivers' licenses for illegal immigrants, although polls showed a huge majority of the state feeling otherwise. In addition, he separated himself from others by proposing new taxes on businesses and the wealthy to balance the anticipated $8 billion deficit for the upcoming fiscal year.[23] This approach clearly differentiated Bustamante from his top-tier opponents, all of whom argued against increasing taxes.

The Bustamante candidacy put those disposed to Gray Davis in an awkward position. While the "No on recall, Yes on Bustamante" approach sounded possible in theory, many feared that those who favored Bustamante would vote "yes" on the recall in order to help Bustamante win the replacement election. In other words, there was real concern that Bustamante represented a threat to Davis. Once Bustamante jumped into the race, leading Democrats argued that the two campaigns were not at odds; but logic suggested otherwise.

Bustamante started out fairly strong. A month into the campaign, a *Los Angeles Times* statewide poll of voters showed Bustamante leading Schwarzenegger by a healthy 35 to 22 percent margin, with the others far behind. A Field Poll at the same time also showed Bustamante defeating Schwarzenegger, but by a smaller margin. A few days later, all thirty-three members of California's Democratic congressional delegation endorsed Bustamante as the replacement candidate should Davis be recalled. By the end of August, Bustamante had secured the endorsement of the California Labor Federation, one of the most powerful interest groups in the state.[24]

Bustamante not only stood out as the only top-tier Democratic replacement candidate but also quickly became a symbol for Latinos, a full one-third of California's population. This was critical, given that Latinos had voted overwhelmingly for the Democrats over the previous decade. Bustamante knew how important it was to secure Latino support, although

75

he fully realized that he needed non-Latinos as well. Straddling the line, at one point he said, "I love my culture. I love everything about it. I love the music, I love the language, I love the food. Look at me, I *really* love the food. But I . . . want to be the governor for everybody in California."[25] Meanwhile, he staked out liberal positions on abortion, the environment, and public education. He also separated himself from others with his "Tough Love for California" theme by proposing to close the state's deficit by raising state income taxes for high-end earners, establishing stricter collection procedures for commercial property taxes, and increasing cigarette taxes.

Then the recall winds shifted. As the campaign entered its last full month in September, Bustamante faced increasing criticism on having received more than $4 million in contributions from Indian casino groups. At issue was not only the amount of money but the claim that Bustamante had circumvented Proposition 34, a 2002 initiative that placed strict limits on how much money a candidate could accept from any single individual or group. After denying any impropriety, Bustamante shifted tactics and gave most of the money to the "No on 54" campaign, an effort to defeat a ballot initiative seeking to eliminate race and ethnicity from public employment records. Not good enough, according to a Sacramento judge, who on September 22 ruled that Bustamante had violated the state's finance law by accepting the funds.[26] Nevertheless, for many the controversy seemed to confirm the belief that Cruz Bustamante represented more of the problem than the solution.

By the last weekend in September, Bustamante was 15 percentage points behind Schwarzenegger in the polls, a complete reversal of the standings a month earlier. Pounded by his opponents, Bustamante seemed to lose more energy with each passing day. Critics described his performance at the many debates as lackluster if not dull. Bustamante's representatives described his behavior as a way to separate himself from the bigger-than-life Schwarzenegger, but it didn't work. As one observer wrote, "Bustamante wanted to be David to Arnold Schwarzenegger's Goliath. But David's aim was true. Too often, Cruz Bustamante's is not."[27] As it turned out, Bustamante's aim was way off the mark on October 7.

Tom McClintock: The Conservative Conscience

Historically, the Republican Party in California has been dogged by a tremendous chasm between conservatives and moderates. At one point that gulf cost the Republicans control of the state assembly even though they were the majority.[28] It has also allowed liberals like U.S. Senator Barbara Boxer to be elected because of terrible Republican infighting during the primaries. Given this contentious background, it should not be surprising that Republicans would face another fracture in the replacement election. This time the challenge to moderate Schwarzenegger came from conservative state senator Tom McClintock.

Unlike most of the 135 candidates, McClintock was no gadfly. He was elected to the assembly in 1992 and then, upon being termed out and after a 2-year hiatus, he was elected to the state senate in 2000. Most impressive, perhaps, was McClintock's showing in his 2002 campaign for state controller against Democrat Steve Westly. Although outspent 5 to 1, McClintock came within 20,000 votes of defeating Westly out of 6.6 million votes cast. Clearly, McClintock stood out as the conservative standard-bearer in 2003. With that in mind, he filed as a replacement candidate.

In his own unique way and style, McClintock wove an unusual tapestry of supporters. Both Indian tribes and evangelicals contributed to his campaign— a combination that seemed possible only in a place like California. Repeatedly, he preached that California's problems lay not in lack of budget revenues but in undisciplined spending and overregulation. Pro-life, antitax, progun, and the only candidate to promise without equivocation that he would not raise taxes under any circumstances, McClintock not only marched to his own drumbeat but brought many others with him. His message was the same, almost word for word, almost everywhere he went, supported by numerous facts and providing incredible consistency.[29] He was the shining light for true believers of the far right.

Throughout the early days of the campaign, polls showed McClintock capturing between 12 and 16 percent of the vote, while Schwarzenegger lagged behind Bustamante. As the weeks went by, Republican leaders began to pressure McClintock to drop out of the campaign. He would hear nothing of it. In fact, the more that key leaders leaned on McClintock to drop out in the name of "party unity," the more he seemed to revel in staying on message. "When I entered this race, I made a promise to stay in right to the finish line. I keep my promises," he said, with less than 2 weeks left in the campaign.[30] That he was outspent by Schwarzenegger by a 10-to-1 ratio was of no consequence to McClintock or his supporters; that he stayed true to his core values was. Nevertheless, McClintock was never able to extend his reach beyond his conservative base.

Peter Camejo: The Liberal Conscience

Peter Camejo secured 5 percent of the vote when he ran for governor as a Green Party candidate in 2002. That a "third party" candidate was able to secure such a vote should have been a warning to Democrat Gray Davis. Most of Camejo's supporters were disaffected Democrats who couldn't bring themselves to vote for Republican Bill Simon, thus Camejo's large protest vote. That success catapulted him into the top tier in 2003, and he was included in the debates.

Portraying both major political parties as "dysfunctional," Camejo offered himself as the true reform candidate. His proposals were as unlikely to draw mainstream support as they were refreshing—a disconnect that he freely acknowledged. Like McClintock, Camejo had a platform, the planks of which he stated with great consistency. All the legislature had to do was raise income taxes on the wealthy and California's budget crisis would be solved, he said. The substantial surplus from the "fair tax" of the wealthy would allow the state to pour more money into public education, solar power, and renewable energy.[31] His sincerity notwithstanding, Camejo's message was not well received by the voters, and his fledgling campaign was quickly relegated to the periphery.

Bill Simon, Peter Ueberroth, and Arianna Huffington: The Dropouts

If leading Republicans in the state were gleeful over a reeling Gray Davis during the summer of 2003, they still had reasons to be concerned. Democrats had only Cruz Bustamante as a top-tier replacement candidate, giving them every opportunity to coalesce by the campaign's end. Republicans, however, were another matter. Gushing over the unexpected opportunity to capture the state's highest office, they lined up for their shot at the prize. And with only a simple plurality required for victory, it seemed that almost anyone could win. Two aspirants who fit this description were Bill Simon and Peter Ueberroth; a third, Arianna Huffington, was a Republican until shortly before the recall campaign. All dropped out before the campaign's end, although their names remained on the ballot.

Bill Simon

In a way, Bill Simon should have been the favored Republican challenger. As his party's standard-bearer just months earlier in the 2002 election, Simon lost to Gray Davis by only 5 percentage points. Although he was outspent 6 to 1 in the 2002 campaign, Simon remained credible to many Republican conservatives. Others, however, viewed the outcome as more of a sign of public disgust with Davis than of any last-minute rally for Simon. Nevertheless, with the public now fully aware of Davis's manipulative efforts in the Republican 2002 primary, Simon plunged into the replacement race.

Almost immediately, Bill Simon found himself the victim of an ideological pincer movement. On one side, he struggled for support from the same

78

Governor Gray Davis rolls up his sleeves to rally supporters in the last days before the election.
(Photo by Jessica Friedman, reprinted courtesy of the South Bay AFL-CIO Labor Council)

Lieutenant Governor Cruz Bustamante (Democrat) holds the floor as Peter Camejo (Green Party) listens. (Photo by Ryan Balbuena)

Independent candidate Arianna Huffington.
(Photo by Bram Goodwin)

Green Party candidate Peter Camejo.
(Photo by Ryan Balbuena)

Republican candidate, and State Senator, Tor
McClintock. (Courtesy of Tom McClintock for State Senat

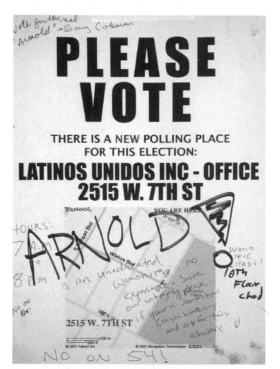

The graffiti on this poster reflects voters' wide-ranging opinions—some humorous, some serious—concerning the recall election.
(Photo by Brian Flemming)

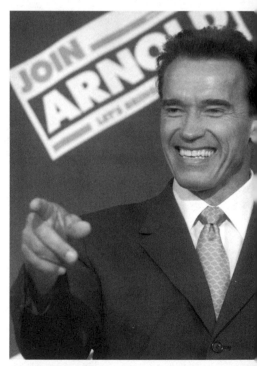

And the winner is . . .
Republican Arnold Schwarzenegger.
(Photo by Ryan Balbuena

pro-life, antitax conservatives courted by Tom McClintock. On the other side, he lost the hope of gaining traction with moderates when, on August 23, the Lincoln Club of Orange County—a Republican grassroots political organization that had been key to Simon's candidacy in 2002—announced its support for Schwarzenegger. The next day Simon pulled out, saying "there are too many Republicans in this race, and the people of our state simply cannot risk a continuation of the Gray Davis legacy."[32] Although Simon eventually endorsed Schwarzenegger, many of his supporters went to McClintock.

Peter Ueberroth

Of the top-tier candidates, Peter Ueberroth stood out as the most one-dimensional. Low key to the point where one sometimes wondered whether he had a pulse, Ueberroth had the reputation of a no-nonsense problem solver. The former major league baseball commissioner and Olympic Games organizer described himself as a "roll up your sleeves and get to work kind of guy" who was more concerned with fixing what was broken than blaming anyone for the problem. "I'm not throwing stones at the Democrats. I'm not throwing stones at the Republicans. I'm throwing stones at the process, which has gotten out of hand and it's got to be fixed,"[33] he said shortly after entering the race.

Pro-choice but a fiscal conservative, Ueberroth positioned himself as a political outsider with the business acumen needed to fix California's ills. But his campaign languished. At the first televised debates with all of the top-tier candidates except Schwarzenegger, Ueberroth answered nearly every question the same way: California needed to become friendly to business again. His monotonic message seemed to reflect his persona, leaving the electorate hungering for more. Stalled with minimal support, Ueberroth dropped out on September 9. Unlike Simon, who endorsed Schwarzenegger, Ueberroth simply faded from the scene. Most of his supporters went to the Schwarzenegger camp.

Arianna Huffington

Of the three top-tier dropouts, Arianna Huffington stood out as the most unique. In 1994, she had campaigned for her then-husband Congressman Michael Huffington in his failed race against Dianne Feinstein to win a seat in the U.S. Senate. Once a staunch conservative Republican, she had become a "born again" liberal who decided to run as an independent. Her new position left Huffington in an awkward place—on the one hand, conservative Republicans no longer trusted her; on the other, liberal Democrats were afraid to trust her. That left Huffington struggling for support from the state's growing number of independents.

Quickly, Huffington seized the left somewhere between Cruz Bustamante and Peter Camejo in California's wide political spectrum. She was the only major candidate who sought repeal of California's Proposition 13, a 1978 measure that reduced property taxes, which had become almost sacrosanct for many Californians. She also defined herself as pro-choice, pro-immigrant, and willing to raise taxes, particularly on business and the wealthy. In the only debate with Schwarzenegger, Huffington accused the actor of being condescending toward women; he told her to relax and "have a decaf." Like the other candidates, the colorful Huffington could not attain any political altitude in the replacement campaign. Unlike the others, however, Huffington announced that she would vote against the recall and campaign full time against Schwarzenegger upon withdrawing on September 30.[34]

The Supporting Cast

Along with the major replacement candidates came 127 others who dreamed they, too, might become governor. Many of the participants were "everyday" people who simply wanted their names on the ballot. Others were people with unusual careers or exotic backgrounds, all of which helped feed the perception of a political circus disguised as an election. Some of the more colorful candidates included the following:

- Iris Adam, Natural Law Party, a business analyst who argued for prevention-oriented natural medicine to prevent disease and save the state money.
- Vik Bajwa, Democrat, a businessman who ran to be the first immigrant to be elected governor; in actuality, that distinction went to Irishman-turned-Californian James Downey, who was elected in 1860.
- Mary "Mary Carey" Cook, Independent, an adult film star who called for a tax on plastic surgery as a way of balancing the state budget.
- Larry Flynt, Democrat, a publisher of adult magazines who ran as "a smut peddler who cares."
- Leo Gallagher, nonpartisan, a comedian who campaigned for a fleet of helicopters to keep the freeways clear of wrecked cars and stalled vehicles.
- Rich Gosse, Republican, who sought "fairness for singles" and promised to decriminalize drugs, gambling, and prostitution to save billions of dollars wasted on arrests, prosecution, and incarceration.
- Garrett Gruener, Democrat, founder of the "Ask Jeeves" search engine, who spent a million dollars of his own money to put forward, mostly on his own website, his ideas for political reform.
- John J. "Jack" Hickey, Libertarian, who promised to do away with public schools because of their excessive cost.

- Gino Martorana, Republican, an Italian restaurant owner who promised a free meal to anyone who voted for him. Not one person accepted.
- Bill Prady, Democrat, a television writer who argued that the voters would be better off with a governor who writes the material rather than one who says the lines.
- Christopher Ranken, Democrat, a planning commissioner who ran as an "anti-recall" candidate and nothing else.
- Reva Renee Renz, Republican, a cocktail lounge owner who promised to void the state's long-term energy contracts and invented a new drink, the Gov Shot (contents unknown).
- James M. Vandeventer, Jr., Republican, whose priorities included granting amnesty to all undocumented workers and legalizing gay unions.

There were others, too, some with thoughtful platforms, some with whimsical ideas. Because of the noise surrounding the large number of top-tier candidates, few attracted any attention. And by the time the number of major candidates had dropped to manageable levels, most people's minds were clearly made up. Nevertheless, the collective character of the remaining replacement candidates put to rest any notion that the recall election would be drab or "politics as usual." The state, the nation, and the world had plenty to talk about.

And the Winner Is . . .

By the campaign's end, the only surprise was the size of Davis's defeat and the extent of Schwarzenegger's victory in the replacement election. Davis was turned out by a margin of 55 to 45 percent, a division that stayed remarkably constant throughout the summer and into fall. The defections were massive and widespread. An NBC exit poll showed that the coalition of past supporters—women, Latinos, union members, and low-income voters—had abandoned Davis in every respect; all dropped in their level of support. Even 25 percent of those who called themselves Democrats voted to recall the Democratic governor.

But the greater surprise lay with the lopsided Schwarzenegger victory. When the votes were counted, the actor turned politician pulled 48.6 percent, with Bustamante a distant second at 31.5 percent; McClintock collected 13.5 percent, while Camejo managed less than 2.8 percent. The remaining 131 candidates shared a meager 3.6 percent of the vote. Schwarzenegger may not have emerged with a resounding mandate, but in a race where some observers initially feared that the winner might need 15 to 18 percent to secure the necessary plurality, he emerged comfortably ahead of the pack. And just like that, the election was over.

7

The Issues

Elections in California have long been more focused on personalities than on issues, and the recall election of 2003 was no exception. The tradition of personality politics in California goes back to Hiram Johnson, the governor who introduced direct democracy. Then, from the 1930s onward, when Democrats had gained a majority among registered voters, Republican candidates for governor still managed to win elections by shifting the focus to personality over party. Republican Earl Warren, most famously, was elected three times. Ronald Reagan won twice. Despite their majority, Democrats elected only four governors in the twentieth century. Two of them were Pat Brown and his son Jerry, both strong personalities.

Yet personality hasn't always been the dominant factor. Republicans George Deukmejian and Pete Wilson were hardly charismatic, but both won two terms in office. Issues played a significant part in their elections, as they did in the elections of Pat and Jerry Brown and others.

Political scientists have asserted that the bad (or good) old days of personality politics in California are over and that the voters have grown more loyally partisan, based on the issue positions of the Democratic and Republican parties.[1]

But the recall election may suggest otherwise. Clearly, the personalities of the candidates were themselves an issue. Gray Davis's lack of personal charisma and leadership skills helped bring his downfall, while Arnold Schwarzenegger's star power made him an instantaneous threat and the ultimate victor.

Yet for all the focus on personality, the voters expressed serious concerns about issues and most of the candidates attempted to address these concerns. Ironically, the ultimate winner was the candidate who was arguably the least specific about traditional issues of public policy. But in addition to these sorts of issues, voters were concerned about the issues of leadership and the way California's political system was working.

82

Table 7.1

Issue Concerns of California Voters

Importance of the candidate's stand	Percent saying "very important"
The state economy	90
Public schools	82
Reducing the budget deficit	82
Expanding business	73
Health care	72
Taxes	71
Crime/law enforcement	62
Electricity supplies, cost	62

Source: The Field Poll, Release #2039, September 10, 2003.

Voter Concerns

We know what voters were worried about from the frequent public opinion surveys reported over the preceding year or two. That list of concerns has been pretty consistent over time.

A September Field Poll asked voters what they considered to be the most important issues in the governor's race. Table 7.1 reports the issues and the percentage that said they were "very important."

A *Los Angeles Times* poll at about the same time came up with the budget deficit, education, the economy, and immigration as the top four concerns. Forty-six percent of the respondents in that poll said "where the candidate stands on issues" was more important to them than the candidate's character.[2] Polls by the Public Policy Institute of California (PPIC) reported the same top three issues, with the recall itself ranking fourth.[3]

In short, as much as some candidates, voters, interest groups, and media wanted to debate abortion, gun control, gay rights, sexual harassment, right-wing coups, and other controversial issues, the voters stayed focused on the essentials and, according to what they told pollsters, made up their minds about candidates accordingly.

Candidate Positions

Contrary to political legend—especially potent in California—candidates did take positions on these and other issues. It's just that their positions on issues were usually not as much fun for the media to report as their lifestyles, mis-

83

Table 7.2

Issue Positions of the Top-Tier Candidates

Issue	Davis	Bustamante	Schwarzenegger	McClintock	Camejo	Huffington
Raise income tax on rich	Y	Y	N	N	Y	Y
Raise cigarette taxes	Y	Y	N	N	Y	N
More cuts in state spending	N	N	Y	Y	Y	N
Abortion: right to choose	Y	Y	Y	N	Y	Y
Rights of domestic partners	Y	Y	Y	N	Y	Y
Death penalty	Y	Y	Y	Y	N	N
Illegal immigrants drivers' licenses	Y	Y	N	N	Y	Y
Reform Prop. 13 (property taxes)	N	N	N	N	Y	Y
Prop. 54 (2003) (racial privacy)	N	N	N	Y	N	N
Prop. 187 (1994) (deny social services to illegal aliens)	N	N	Y	Y	N	N

Source: "Recall Notebook," prepared by Larry N. Gerston for NBC 11, San Jose, October 7, 2003.

statements, and peccadilloes. Table 7.2 reports the positions taken by the major candidates on ten issues.

Candidates expressed their positions on these and other issues in a variety of ways. Governor Davis made his positions on many issues known by the bills he signed into law while the recall was in process, but his campaign also attempted to make the recall itself an issue by labeling it a right-wing coup. Davis and his supporters attempted to tout his record, which, while substantial, gained him little advantage in his final campaign.

Among the replacement candidates, several, including Cruz Bustamante and Arnold Schwarzenegger, held press conferences on such issues as the budget and political reform, and some issued position papers. Besides Davis, Tom McClintock probably had the most specific positions on the widest range of issues by virtue of serving in the legislature but also because he is a smart and thoughtful if somewhat dogmatic conservative. On the other end of the political spectrum, Peter Camejo articulated liberal policies and proposals intelligently. Arianna Huffington's positions on the issues were also for the most part liberal, although the breadth and depth of her knowledge were more limited. She was particularly outspoken, however, on reforms of campaigns and taxation.

Overall, Camejo, Davis, and McClintock could be said to have been most specific on issues, but all the candidates took their obligation to express their views reasonably seriously, largely because the press—through questions at debates, press conferences, and on the campaign trail—would not let them do otherwise. But Schwarzenegger, because he avoided debates and direct questions from political reporters, was probably the least specific on the issues.

The Economy and the Budget

The economy and the budget were the foremost issues for the public, the media, and—whether they liked it or not—the candidates. Gray Davis was faulted for his leadership, but without the recession, the huge budget deficit, and the political deadlock over the previous two budgets, his leadership failures might not have been of such concern. As unemployment rose and tax revenues fell, the dearth of leadership in Sacramento became salient.

To some extent, Davis was a scapegoat on these issues. Most observers close to the process agree that the budget the legislature ultimately approved in 2003 was worse than the one Davis proposed. While Davis had balanced significant spending cuts with selected tax increases, the budget he ultimately signed was a jerrybuilt package of cuts and borrowing that many believed deferred rather than solved the state's fiscal problems. The legislature may have deserved more blame for this than Davis did, yet his failure to lead—if any governor could have effectively led the legislature—was apparent. And the recall petitions were circulating just as public anger boiled over.

The vehicle license fee or "car tax" was one element of the budget package that especially inflamed public opinion. In the previous budget crisis, when Republican Pete Wilson was governor, the car tax was raised to balance the budget and fund city and county services, including police and fire protection. In 1998, when the economy had recovered, Governor Wilson and the legislature agreed to cut the fee by two-thirds on condition that the state would continue to provide other funds for cities and counties. The legislation stipulated that the fees would rise again if the state couldn't supply such funds. In June, with the deficit spiraling out of control, Davis's director of finance ruled that such a time had come and ordered higher fees reinstated, tripling the existing rates. Although Republicans objected, the $4 billion the fees would provide were essential to the ultimate budget deal.

The new rates weren't supposed to go into effect until October 1, but the reaction was immediate and intense. Conservative leaders stirred the controversy, but in car-happy California, citizens across the political spectrum were appalled to learn that their fees would increase, in some cases by hundreds of dollars. The decision was raw meat for talk radio, where it was redefined as the "car tax" rather than the "vehicle license fee." Davis was blamed and the announcement came just as Darrell Issa was injecting cash into the recall effort. Angry voters (and drivers) lined up to sign petitions.

"It's a tax everyone understands," declared conservative talk radio host Roger Hedgecock. "It isn't so far below the radar, like the sales tax is, that it gets folded in and you don't think about it; or the gas tax. You get a bill for $600 and you go, 'What?' It hit people like a wet fish in the face."[4] Not surprisingly, 58 percent of Californians opposed reinstating the vehicle license fee in a June PPIC poll.[5] The numbers were higher among supporters of Republican candidates Schwarzenegger and McClintock, both of whom promised to rescind the tax, while Bustamante said he'd reduce it—and Davis was stuck with it.

But while the vehicle license fee was a lightning-rod tax issue, voters had more general concerns about taxation and wasteful government spending: taxes are too high in California and government could spend less. Even as the governor and the legislature were finally agreeing on a budget, polls reported that 70 percent of Californians thought "the state government could spend less and still provide the same level of services."[6]

Candidates McClintock and Schwarzenegger agreed. Both promised to cut wasteful government spending, rescind the car tax, and avoid other tax increases. McClintock, the true conservative, pledged vigorous spending cuts as well as cuts in existing taxes and no new taxes of any form. Schwarzenegger, the moderate, also rejected the car tax and promised spending cuts; but while he turned up the antitax rhetoric, he was unwilling to pledge no new taxes ever.

"I feel the people of California have been punished enough," declared Schwarzenegger. "From the time they get up in the morning and flush the toilet, they're taxed. When they go get a coffee, they're taxed. When they get in their car, they're taxed. When they go to the gas station, they're taxed. When they go to lunch, they're taxed. This goes on all day long. Tax. Tax. Tax. Tax. Tax."[7] Schwarzenegger said he would appoint an auditor to review state spending and uncover waste, at the same time promising not to cut spending on education (54 percent of the state budget). Despite the intense concern of voters about taxes and the budget, Schwarzenegger avoided specific positions on taxation, surrounding himself with prominent conservative economists and business leaders to send a symbolic message suggesting his concerns on this issue. He did, however, proclaim his support for Proposition 13, the initiative that limited property tax increases, after financier and Schwarzenegger adviser Warren Buffett publicly pointed out its flaws— sending Buffet off to do "500 sit-ups."[8]

Schwarzenegger's long-term solution to the state's economic problems was to reduce state regulations and taxes on businesses to create more jobs. "I will stand up to anti-business legislation, I will stand up to Sacramento's job-killing philosophy," he told the California Chamber of Commerce.[9] California's flawed workmen's compensation program was a special (and justified) target for Schwarzenegger and his business supporters, although voters were unconcerned. They did, however, seem to agree about improving the business climate to generate jobs, rating the economy and jobs as the state's top problem in the September 2003 PPIC Statewide Survey.[10]

Other candidates took less popular stands. Unlike every other top-tier candidate, Arianna Huffington proposed reform of Proposition 13 to increase taxes on commercial and industrial property, and she advocated the elimination of tax shelters and loopholes that benefit corporations and the rich—not exactly "business-friendly." Peter Camejo also supported reform of commercial property tax assessments, along with an increase in the tax rate for the wealthy. Democrat Bustamante advocated a similar increase in taxation for top earners, as well as closing loopholes, increasing alcohol and tobacco taxes, and pursuing tax evaders. Although Governor Davis was stuck with the budget he signed, he supported increases in the tobacco tax and higher income taxes on the rich to replace funds lost from the vehicle license fee. He also proposed a blue-ribbon panel to consider structural reforms in the budget. All four of these candidates also promised further cuts in spending in various ways.

Public opinion polls all suggest that McClintock and Schwarzenegger were taking the more popular positions on these issues, yet state taxes and spending may not have been as bad in reality as in the public perception. The media reported, for example, that California ranked eighth among the fifty states on per capita taxes and eighteenth on taxes as a percent of income. On both measures California was above the national average, but higher costs and more services could account for that. The overall state and local tax burden in California is lower for most families than the national average, although it's higher for the top 15 percent.[11] A study by the Rockefeller Institute of Government ranked California nineteenth of the fifty states in the tax burden on individuals and businesses.[12] Whatever the reality, the voters preferred to believe they could have the services they wanted while cutting spending and avoiding tax increases.

Education: We're All for the Kids

Education has long ranked at the top of the list of voter concerns. It's also the top spending category for California state government. Yet education was scarcely given lip service in the recall campaign of 2003. All the major candidates proclaimed their support for education. Bustamante, Huffington, and Schwarzenegger promised not to cut K–12 education. Davis promised to maintain school funding but warned that cuts could be necessary. Only McClintock was confident that cuts in education were possible while still putting more funds into classrooms.

Schwarzenegger enhanced his moderate image by his opposition to vouchers (along with Bustamante, Camejo, Huffington, and Davis), but conservatives (and others) appreciated his attack on Sacramento as a "schoolyard bully" for overregulating local schools.[13]

Education should have been a good issue for Gray Davis. Responding to voter concerns, he declared it his top priority when he took office in 1999, and education spending increased by $7.3 billion between 1999 and 2004. Spending per pupil rose $1,000, improving California's national ranking in spending per student from forty-third to twenty-sixth. Classes were smaller and test scores rose. Yet Davis gained little public support from these improvements. In the public perception, education was still a top problem and Gray Davis had not fixed it.

Higher education was discussed even less than K–12, although California provides one of the best and most affordable public systems of higher education in the world. Here, too, the Davis administration could boast of holding tuition and fees steady for years while increasing aid to needy students. But the massive deficit of 2003 resulted in higher fees—just in time for students and parents to vent their frustration about rising education bills in the recall election. Bustamante, Camejo, Huffington, and McClintock all advocated rolling back fee increases. Schwarzenegger blamed the increases on "mismanagement in Sacramento" and pledged to create a more stable system, but he stopped short of promising a rollback. He also asked for improved "efficiencies" at institutions of higher learning.[14]

Energy: Crisis and Conspiracy Theories

Californians have ranked energy supply and utility costs among their top concerns in public opinion polls since the energy crisis started in 2000 and climaxed in 2001. These issues remained on the minds of voters in the recall election but not at the forefront of debate. Rather, they formed a sort of back-story to the recall, having shaped perceptions of Gray Davis as a lackadaisi-cal and ineffective leader, responding too late and too weakly to rapidly rising prices, shortages, and blackouts. Support for the recall—as demonstrated in polls and in numbers of signatures on petitions—was strongest in San Diego and other counties where the energy crisis hit first.

Davis defended his record, pointing out that the energy crisis had been resolved, albeit expensively. He blamed the federal government (and Bush administration) for failing to intervene to lower the costs of contracts Cali-fornia negotiated with energy suppliers, also noting his administration's ef-forts to recoup some of the costs through the courts. Perhaps more substantively, Davis boasted that during his time in office thirty-seven new power plants had been licensed and more were in the pipeline, while none had been licensed between 1994 and 1998. As Davis liked to point out, Re-publican Pete Wilson was governor during those years, and it was Wilson who signed the legislation that deregulated the utilities—and who co-chaired Arnold Schwarzenegger's campaign. Try as he did, however, Davis was never able to shift the blame for the crisis to his predecessor. He was in office when the crisis hit and the public held him responsible.

Schwarzenegger took advantage of this and attacked Davis for his han-dling of the crisis, promising to facilitate the building of new power plants and renegotiate Davis's expensive contracts with energy suppliers. His posi-tion on energy issues was tainted, however, by the revelation late in the cam-paign that Schwarzenegger had met privately in 2001 with Ken Lay of the now discredited energy company Enron. At the time, Ken Lay and Enron were under threat of a massive lawsuit demanding a refund of $9 billion to the state of California. Governor Davis was unwilling to settle the lawsuit. Online reporter Greg Palast reported the meeting as the beginning of the conspiracy to recall Gray Davis —with Schwarzenegger as his intended suc-cessor.[15] On his election, Schwarzenegger promised to revisit energy de-regulation, much to the chagrin of consumer advocates.

Cruz Bustamante, meanwhile, criticized the energy companies throughout the campaign and boasted that he had brought a $9 billion dollar lawsuit against Enron, one of the companies accused of price manipulation during the energy crisis. His position distinguished him from Gray Davis, but not enough to mat-ter to voters angered by steep increases in their monthly utility bills.

Immigration—and Immigrant Candidates

Immigration has been an issue in California politics throughout history, some-times in the forefront, but always lurking somewhere in the state's political consciousness. Ironically for a state where everyone but Native Americans has immigrant roots, anti-immigrant attitudes are strong. As of 2000, a total of 26.2 percent of California's population was foreign-born—and millions more were first-generation Americans. Yet immigration has ranked high on the list of voter concerns. In 1994, Republican Governor Pete Wilson, suffering from approval ratings almost as low as those of Gray Davis in 2001, exploited anti-immigrant sentiments to win reelection by tying his campaign to Proposition 187, a popular initiative that reduced government benefits to illegal immigrants.

Proposition 187 remained controversial nearly ten years later and, predictably, the issue came up in the recall election. Davis, Bustamante, Camejo, and Huffington opposed Proposition 187 and the Davis administration— belatedly according to Bustamante—settled the lawsuit on the measure. Tom McClintock forthrightly supported 187 in 1994 and 2003. Arnold Schwarzenegger, himself an immigrant, admitted to having voted for Proposition 187, blaming federal immigration policy for California's problems.

But the galvanizing issue on immigration in 2003 was a bill passed by the legislature and signed by Governor Davis to allow illegal immigrants to obtain California drivers' licenses. Davis had previously vetoed similar legislation, but under pressure from a liberal majority in the legislature and eager to appease Latino voters, he signed the 2003 law. Schwarzenegger and McClintock opposed the legislation because it did not provide security checks in an era of terrorism, promising to repeal the legislation if elected. The Field Poll reported 59 percent of the voters also opposed the policy,[16] and like the vehicle license fee, the driver's license issue became raw meat for talk radio, further riling up many angry voters.

Another issue on the ballot in 2003 related to race and immigration policy. Proposition 54, the "racial privacy" initiative, would have banned California state and local governments from collecting data on race and ethnicity. Businessman and University of California Regent Ward Connerly, who had successfully sponsored a 1996 proposition banning affirmative action by California state and local governments, was the chief sponsor and advocate for Proposition 54, arguing that labeling people by race and ethnicity was a hindrance to the advancement of minorities. Leaders of California's racial minority groups, however, perceived Proposition 54 as an attack that would blind state and local governments to the unequal treatment and needs of their constituencies, and campaigned vigorously against the measure.

All the candidates except Tom McClintock expressed their opposition to Proposition 54 and, on election day, 64 percent of the voters rejected it. Connerly blamed the recall for diverting attention from the issue.

Opponents of the recall and supporters of Bustamante hoped to use Proposition 54 to rally their troops, but opposition to 54 did not carry over to support for either Davis or Bustamante. In fact, Schwarzenegger may have been the major beneficiary of voter attitudes on the issue. His opposition to the measure reassured many voters that he was a moderate on these issues, while his own status as an immigrant gave him credibility.

Voters were reminded of that status as Schwarzenegger told his own inspiring story, but they were also influenced by two issues that arose relating to his immigrant status. In an aside to a supporter, Governor Davis made fun of Schwarzenegger's pronunciation of "California." "You shouldn't be governor unless you can pronounce the name of the state," Davis said.[17] Davis was immediately attacked for his insensitivity to the state's many immigrants, and Peter Camejo pointed out that the governor's own anglicized pronunciation of the name of the state was a perversion of proper Spanish pronunciation. The state senate, with many Democrats abstaining, passed a resolution demanding that Davis apologize for his remark. Davis said he was "just kidding," and issued a sort of apology.[18]

Schwarzenegger's immigrant status also came up when questions arose about his visa status when he first came to America. Allegations were made that he had violated his visa by accepting work. No solid evidence emerged, however, and the issue gave Schwarzenegger another opening to tell his story and advocate a federal policy to provide appropriate work permits for immigrants.

Historically, issues relating to race, ethnicity, and immigration have worked to the advantage of Democratic candidates, as Republicans were seen to be anti-immigrant and anti-minority. A series of ballot measures established this in the minds of many Latino, African-American, and Asian voters (with frequent reminders from Democrats) and drove them to register and vote Democratic. Had this held true in 2003, Davis might not have been recalled—or Bustamante would have become governor. But it did not. Schwarzenegger's position on Proposition 54 and his own story as an immigrant appealed to many of similar status and turned the tide among minority voters away from the Democrats—or at least stemmed the flow.

Social Issues: Gay Rights, Gun Control, Abortion, and Women's Rights

As with race, ethnicity, and immigration issues, social issues in California have historically swung women and gay and lesbian voters toward Democratic candidates, often by large margins. Gray Davis could boast a strong record on gun control and a woman's right to choose and even greater progress during his tenure on gay rights, with the signing of two bills significantly increasing the rights of domestic partners. All of the candidates on the left—Davis, Bustamante, Camejo, and Huffington—lined up solidly in favor of gay rights, gun control, and abortion rights. On the right, McClintock opposed all three.

That left the center for Schwarzenegger, who, unlike most California Republican candidates, supported all three, albeit with qualifications. He supported gay rights but not gay marriage. He supported abortion rights, but not "partial birth" abortion. And he supported some forms of gun control. His position on these issues added to his credentials as a Republican moderate and attracted many mainstream voters, including women, gays, and lesbians.

His success with these constituencies was seriously threatened as election day approached by articles in the *Los Angeles Times*—soon printed and broadcast around the state, the nation, and the world—alleging that he had sexually harassed a number of women, fondling their breasts and touching them inappropriately, usually on movie sets. Such a bombshell in the last days of a campaign can often destroy a candidate's chances. This time it didn't. Schwarzenegger, perhaps learning from the sad experience of President Clinton, immediately apologized while not quite admitting guilt. "I want to say to you, yes, that I have behaved badly sometimes. Yes, it is true that I was on rowdy movies sets and I have done things that were not right, which I thought then was playful but now I recognize that I offended people. Those people that I have offended, I want to say to them I am deeply sorry about that and I apologize because that's not what I'm trying to do."[19] Significantly, Schwarzenegger's wife, Maria Shriver, a member of the Kennedy clan, rallied to his side, defending him and increasing the number of her public appearances.

Shriver had become more visible in the campaign earlier, when other concerns about her husband's treatment of women were raised. Newspapers reported an interview with *Oui* magazine in which Schwarzenegger seemed to boast of group sex. Remarks about forbidding his wife to wear pants in public were reported and seen as sexist. Even characters he played in films were cited as evidence of sexism. Women's groups regularly picketed his public appearances. On one such occasion, they held up toilet seats with "Flush Arnold" signs, a reference to a scene in *Terminator 3* in which Schwarzenegger shoves the head of a female robot into a toilet.

But Schwarzenegger's star quality, quick apology, and the support of Maria Shriver seem to have prevented these charges from doing his candidacy serious damage. He blamed Gray Davis for the allegations in the *Los Angeles Times*, accusing Davis of running another "puke campaign."[20] Davis denied responsibility, but his reputation as a dirty campaigner limited the credibility of his denial. In fact, the stories on sexual harassment were the work of the

Times itself—leading some to believe that the state's leading newspaper was out to get Schwarzenegger. Similarly, accusations of anti-Semitism and reminders of his Austrian roots flared up and faded out. In the end, these charges may have hurt Schwarzenegger, but not enough to deny him victory.

Other Issues: The Environment, Health Care, and Political Reform

Other issues were also discussed, but not as much as these. The environment, always a concern of Californians, usually works to the advantage of Democrats as an issue, and here again, Davis could cite a strong record, including recent trend-setting legislation on global warming. But Schwarzenegger neutralized the issue by proclaiming himself environment-friendly and taking positions on many environmental issues that other Republican candidates—including Tom McClintock—would not. The Democrats again failed to gain traction on one of their core issues.

As the campaign heated up—and as it became increasingly apparent that Gray Davis's days in office were numbered—the liberal majority in the state legislature got busy, passing a backlog of liberal bills in hopes that Davis would sign them under electoral pressure and knowing that Governor Schwarzenegger surely would not. These included the driver's license bill, landmark legislation to protect the privacy of financial data, and others, but most prominent among them was Senate Bill (SB) 2, a bill to provide health care to many of the 7 million uninsured Californians through an employment-based program. Business leaders opposed the bill because it added to their costs. Labor unions passionately supported it. Absent the recall, the cautious Davis would almost certainly have vetoed SB 2. But with unions leading the opposition to his recall, he was under heavy pressure to sign the bill. Keeping both labor supporters and business opponents on tenterhooks for days, Davis signed the bill just 2 days before the election. While Democrats and liberals supported SB 2, Schwarzenegger and McClintock opposed it and promised repeal. Few issues are as important to Californians as health care, but even this landmark legislation was not enough to save Gray Davis—or even merit much discussion in the campaign.

Political reform, an almost constant issue in California politics since the days of Hiram Johnson, was also a campaign issue. Davis's pay-to-play reputation, demanding donations before he would consider policies or appointments favorable to the potential donors, was used against him. Similarly, Cruz Bustamante's candidacy was seriously damaged by what some perceived as questionable contributions from Indian gaming interests and public employee unions.

While all the candidates paid lip service to campaign reform, Arianna Huffington was most outspoken about it, declaring "We should recall the system"[21] and promising to sponsor a campaign reform initiative. Schwarzenegger, too, was forthright on reform, announcing a plan for "political openness," including a ban on fund-raising while the budget is being negotiated, a panel of judges to oversee legislative redistricting, and a promise to veto any legislation that does not receive full and public debate in the legislature. While few voters may have made their minds up on these issues, Schwarzenegger surely scored points with his positions, distinguishing himself from pay-to-play and politics as usual and bolstering his credentials as a moderate.

Issues: The Bottom Line

All these issues and more were discussed and debated. The economy and the budget seemed of most intense interest, followed by the vehicle license fee. Other issues—from education and the environment to drivers' licenses for illegal immigrants, the rights of domestic partners, and a woman's right to choose—played to particular constituencies and may have affected their votes. Overall, the candidates in the recall election, including Arnold Schwarzenegger, deserve credit for staking out positions on a wide range of issues. While it has been said that California politics is more about personalities than issues, these candidates could not be faulted for campaigning on personality alone.

Yet in the end, the race—and the issues—may have come down to personality or, rather, leadership. Gray Davis and Cruz Bustamante, lacking in charisma, campaigned on experience and a record of accomplishments. That wasn't enough for either of them. Tom McClintock became the hero of California's conservatives—even as his candidacy threatened to siphon votes away from Schwarzenegger and possibly prevent a Republican victory. Arnold Schwarzenegger, attacked for his lack of experience in government and politics, at first seemed at a loss on specifics. He made misstatements on issues throughout the campaign, beginning with a promise "not to take money from any one,"[22] then apologizing when he began fund-raising with business donors. Later, he rashly announced and then swiftly retracted his intention to close the state Environmental Protection Agency.[23] Schwarzenegger avoided the press so as to prevent more mistakes, but he also presented detailed proposals on the economy, the environment, and political reform, attempting to compensate for his lack of a record.

According to a *Los Angeles Times* survey, 46 percent of the voters said the candidates' stand on issues was most important to them in deciding for whom to vote. But 30 percent said "character and leadership ability" were most important and another 22 percent said both character and positions on issues were important. Sixty-six percent said no single issue would determine their vote.[24] Specific positions on issues may have mattered to interest groups and other campaign contributors and surely affected many votes, but for most voters, the issue was leadership—and change.

As Arnold Schwarzenegger declared in one of his television ads, "It comes down to this: If you are happy with the way things are, then keep your current leaders. If you want to change this state, then join me."[25]

8

Interest Groups

Taking Sides

None of the other states and few nations are as diverse as California, where diversity seems to be layered on diversity. From race and ethnicity to income and class; from varied economic interests to competing political ideologies; from geographic region to rural, suburban, and urban interests within regions; from labor unions to lawyers to Indian casinos—diversity is a joyous if sometimes fractious fact of life in California. And every element of that diversity is expressed through organized interest groups.

Throughout the state's history, all these interests and many more have been actively engaged in government and politics. More than in most other states, they are virtual partners in legislation and elections in California. Through their active presence in Sacramento, they lobby, cajole, and persuade, shaping the state's laws. During election years, they provide money and volunteers to candidates who share their views. If they can't get their way through the legislative process, they take their case directly to the voters through the initiative process. Indeed, they have become the primary source of ballot measures. And when all else fails, they take their cases to court.

The weak organizational structure of California's political parties enhances the power of interest groups and the dependence of candidates on their money. Term limits have made them the senior and most knowledgeable denizens of the state capitol. Some would say it's not our elected officials but these groups and their lobbyists that are really running California. Looking back at the "perfect storm" that precipitated the recall, it's easy to see that interest-group politics had a lot to do with the energy crisis, running up the budget deficit, and even the election of Governor "pay-to-play" Davis. Perhaps it is they who should have been recalled.

Inevitably, many such groups were involved in the recall. But were they as great a force in the recall election as they are in regular elections? Or did the constricted time frame and the open ballot lessen their power somewhat?

The victor in the recall explicitly campaigned against these "special interests," which seemed to appeal to many voters. But was Arnold Schwarzenegger free of their influence? Or was he, like many of us, merely condemning as "special interests" groups other than those supporting him?

The Right: Conservative Groups and the Recall

Conservative groups of various sorts were among the first to climb onto the recall bandwagon. While they didn't always agree on a replacement candidate, they were unanimous in their enthusiasm for recalling Gray Davis.

The People's Advocate, led by Ted Costa, was the first interest group to engage in recall politics. According to Mark Abernathy, a political consultant and an ally of Costa's, they considered the simultaneous recall and winner-take-all replacement election an ideal opportunity for a candidate like Arnold Schwarzenegger right from the start.[1] Costa, however, was careful to keep the focus throughout on recalling Gray Davis rather than on specific alternative candidates.

However, the Howard Jarvis Taxpayers Association, another antitax, antigovernment group, did not take on the recall as a primary cause. Jon Coupal and other leaders of the Jarvis group nevertheless contributed to recall fervor by their ongoing critique of California's wasteful government spending and high taxes—particularly the car tax. Once the recall qualified, however, the group endorsed Schwarzenegger. The imprimatur of the Howard Jarvis Taxpayers Association gave him instant credibility with antitax conservatives and was eagerly sought after and highly valued by his campaign.[2]

Other conservative groups piled on. Churches aren't normally considered interest groups, but when they get involved in politics over issues such as abortion or gay rights, they take on the characteristics of interest groups. Outraged by California's liberal abortion and gay rights policies, religious conservatives—whether through their churches or other groups—enthusiastically supported the recall. Religious conservatives and antitax activists in the Republican Party led the drive for the state party convention to support the recall as early as February—even as the party leadership held back, at least until the Lincoln Club of Orange County endorsed the recall in May. Other party clubs and conservative groups followed. The National Rifle Association jumped at the chance to get rid of Gray Davis, who had signed several gun-control bills.

Once the recall qualified for the ballot, these conservative groups were in something of a quandary, however. Arnold Schwarzenegger was the Republican candidate best positioned to win, but while he presented himself as an antitax, antigovernment candidate, he was a moderate on issues crucial to these staunchly conservative groups, including abortion, gay rights, and even gun control. In the eyes of the conservative right, Tom McClintock was the only true conservative in the race. His positions on issues were forthrightly conservative, and he asserted them with confidence and intellectual aplomb. But could he win? With McClintock never having run better than third in the polls or gained the support of even 20 percent of the voters, conservatives had to consider compromising their principles to vote for an imperfect but more likely winner—Arnold Schwarzenegger—or risk a victory for Democrat Cruz Bustamante. While Republican Party leaders across the political spectrum flocked to the more likely winner, more ideological conservatives and their organizations stuck with the purer McClintock.

The Left: Liberal Groups and the Recall

Similarly, liberal groups lined up against the recall and, for the most part, in support of Cruz Bustamante as a replacement candidate should the recall succeed.

The Sierra Club, the League of Conservation Voters, and other environmental groups were solidly opposed to the recall, not only because Gray Davis had a strong record on the environment but also out of fear that a Republican governor might reverse policies they'd worked so hard to initiate. Schwarzenegger's moderate position on environmental issues, however, reduced this fear—at least until he misspoke at a meeting with farmers in Fresno and said he would get rid of the state's Environmental Protection Agency if elected governor. His campaign declared that he had no such plan.

Like environmentalists, gay and lesbian groups opposed the recall and supported Bustamante. Davis signed two major domestic partners bills over the strong objections of Republicans and religious conservatives, but California voters had also approved an amendment to the state constitution defining marriage as between a man and a woman during Davis's time in office. Gay and lesbian leaders viewed their progress on domestic rights as tenuous at best; they feared a reversal if a Republican gained the office of governor. As on environmental issues, however, Schwarzenegger's moderate position on gay rights made him at least a possibly acceptable candidate for gays and lesbians.

Women's groups were just as concerned about a possible reversal of progressive policies for women—especially on abortion; in the end, they were less likely to see Schwarzenegger as an acceptable moderate alternative. Feminist groups both within and outside the California Democratic Party have long been a mainstay of the state's liberal establishment. They've helped elect ever more women to the state legislature and higher office as well, including California's two U.S. senators. They've campaigned and lobbied for women's issues but also health care, child care, education, gun control, and more. In the recall election, women's groups such as the California Chapter of the National Organization for Women (NOW) stuck with Gray Davis. Some women's groups supported Cruz Bustamante as a replacement candidate because he was seen as the strongest candidate. Arianna Huffington, an assertive feminist herself, was attractive to some women activists, but others worried about her late conversion to liberalism and more felt that the best defense of their policy victories was to keep Gray Davis in office.

Feminist leaders criticized Arnold Schwarzenegger throughout his campaign, upset by allegations of his past sexist behavior and remarks he made himself in earlier interviews as well as by the purported sexism of characters the actor portrayed in his films. Some activist women's groups staged loud, sometimes rowdy protests outside many Schwarzenegger events, harassing the harasser in their view. The *Los Angeles Times* stories about Schwarzenegger's alleged fondling of female coworkers further inflamed feminists, and leaders of various women's groups figured prominently in commentary in and on the *Times* stories. Schwarzenegger countered with public appearances by his wife, Maria Shriver. A pro-choice group called Women for Arnold was also formed and raised $102,000 to advocate for their candidate.

African Americans are another core liberal constituency in California. Through organizations like the National Association for the Advancement of Colored People (NAACP) and even more often through their churches, African-American leaders keep their members informed and engaged in politics—almost always on the side of liberals and Democrats. The recall was no exception. Leaders from Willie Brown to Danny Glover lined up in support of Gray Davis, along with key organizations and church leaders. Noted civil rights leader Jesse Jackson used his energy to mobilize African Americans in the Los Angeles area, where their numbers are greatest. Cruz Bustamante was their preferred replacement candidate, even though in the past he had offended blacks when the "*n* word" slipped out while he was talking to a group of black labor leaders—a slip of the tongue, he said in his swift apology.

Latinos—mostly Mexican American in California—are a more recent and less solidified component of the liberal Democratic coalition. Despite their significant numbers, Latinos have only recently become a political force in California, constituting a rising proportion of the voters in every election for the past decade or more. Given their Roman Catholic background and strong family orientation, Latinos can be somewhat conservative on social issues such as abortion and gay rights; but for the most part they are liberal on economic issues and government services. This complexity might have made

Latinos fair game for both major political parties, but by endorsing a series of initiatives that were perceived as anti-immigrant and anti-minority (Propositions 187, 209, 227, 54), the Republican Party drove Latinos into the eagerly open arms of Democrats in the 1990s.

Latino groups and their leaders found themselves in an even more complicated situation in the recall election. Gray Davis had signed legislation, albeit sometimes reluctantly, of great importance to Latino groups like the United Farm Workers. They owed him something. On the other hand, with Cruz Bustamante as the sole major Democrat on the replacement ballot, they had a tempting opportunity to elect a Latino governor of California for the first time since 1871. Nevertheless, most Latino groups and their leaders dutifully opposed the recall while supporting Bustamante. The real question was where would they put their energy. Davis reached out to them by signing the controversial driver's license bill. Bustamante emphasized his roots and worked his Latino constituency hard. Republican attacks on Bustamante for his membership as a student in Movimiento Estudiantil Chicano de Aztlán (MEChA), because of that organization's radical manifesto, may have helped him with Latino leaders and voters. The manifesto, written in the 1960s, did sound pretty radical, with rhetoric about "brutal gringo invasion of our territories" and the need for "self-defense against the occupying forces of the oppressor at every school"[3]; but MEChA today is well known on college campuses in California and well within the mainstream. On the other hand, Bustamante hurt himself with Latino voters when he failed to win the endorsement of the Mexican American Political Association (MAPA), the leading statewide political organization for Latinos, because a scheduling mixup made him late to their meeting.

Just in case any of these groups forgot whom they were supposed to support, Governor Davis signed bills intended to please all of them—and other groups as well. For environmentalists, there was legislation on greenhouse gases and the recycling of computer monitors; for gays and lesbians, domestic partners rights; for Latinos, drivers' licenses for illegal immigrants. For minorities in general, Davis announced his opposition to Proposition 54, the ballot measure to ban the collection of ethnic and racial data by the state. For consumers, there was auto insurance reform and a privacy bill. For Indian tribes, protections for sacred grounds. For business, a modest reform of workers' compensation insurance. And for labor there was even more.

Labor: Rallying the Troops

No group is more important to the liberal Democratic coalition in California than labor, not only in number of voters but also, and perhaps more significantly, for money and volunteers. That may seem surprising. California doesn't leap to mind as a "union" state. But in fact, while labor union membership has been shrinking across the country, it has grown in California. Nationally, 13.3 percent of all workers are union members, while 17.8 percent of California's workers belong to unions—a total of 2 million members.

California has a long and proud tradition of union activism,[4] but most people know about neither that history nor the current strength of unions in the state. Union membership in manufacturing industries has declined in California as these industries have declined. But union membership in the public sector and service industries has grown exponentially. Public employees, teachers, health care workers, janitors, and others have been joining unions in large numbers in recent decades, more than replacing declining membership in manufacturing. Most of these workers are affected far more significantly by public policy than workers in traditional manufacturing industries, so their unions are particularly active in elections and lobbying.

Unions—from the building trades to teachers, prison guards, and other public employee groups—have long supported Gray Davis. They campaigned hard for him in both his gubernatorial campaigns and benefited accordingly. Prison guards, major financial supporters of Davis's campaigns, received big pay raises. Other public employee unions did better under Davis than under his Republican predecessors, although they weren't always satisfied with what they got. Private sector unions were delighted by Davis's decision to strengthen prevailing wage laws and reinstate requirements for overtime pay for workdays longer than 8 hours. He also, somewhat reluctantly, signed legislation to accelerate collective bargaining for agricultural workers. Unions were sometimes frustrated with Davis's caution and incrementalism, but these actions and others rallied them to his defense in the recall election.

When the anti-recall committee was formed, unions were at its core and a member of the firefighters' union was chosen as the committee's chair and public face. Coincidentally, while the campaign was under way, firefighters and police unions were lobbying for Davis to sign a law giving their organizations the right to binding arbitration in disputes with their local government employers. Davis signed the bill shortly after the election.

Art Pulaski, leader of the statewide AFL-CIO (an umbrella organization for virtually all the unions in California), committed his organization and all its unions to fight the recall. Governor Davis went to a national AFL-CIO meeting to ask for help and was promised that it would be forthcoming.

It was. Unions spent nearly $10 million on the campaign against the recall. Individual unions and county and statewide organizations contributed to anti-recall committees and Bustamante's campaign, also spending money independently to oppose the recall as well as Proposition 54. Most union members throughout the state received multiple mailings from the state labor federation, urging "Vote NO on the Recall. Vote FOR Cruz Bustamante. Vote NO on Proposition 54." A final mailing featured a full copy of the ballot that voters would see on Election Day, with the union-recommended way to vote highlighted. Their goal was to help voters through an unusually complicated ballot listing all 135 candidates for governor. Because the placement of candidate names varied from district to district, multiple versions of the ballot were necessary for the mailing.

Individual unions gave money and campaigned among their members. The California Teachers Association, for example, announced its opposition to the recall as early as May 31 and went on to give $125,000 to anti-recall efforts. The California State Association of Electrical Workers gave $160,000 and the Professional Engineers of California gave $726,200 (mostly to Bustamante).

Meanwhile, as the statewide Labor Federation focused on campaign contributions and mailing, local labor councils were organizing phone banks and door-to-door campaigning. These efforts varied from county to county. Some local labor councils—including those of Los Angeles, Santa Clara, and San Diego Counties—have developed sophisticated and energetic volunteer organizations; others have been less successful. But with so much at stake, union activity was intense in many counties. Labor leaders claimed that, statewide, union volunteers made over 35,000 calls asking people to vote "No on Recall, Yes on Bustamante," and no on Proposition 54. "Our message to our members," said Miguel Contreras, leader of the Los Angeles County Labor Federation, "is voting for Arnold Schwarzenegger is like a chicken voting for Colonel Sanders."[5]

But despite the money and the hard work of many, some unions were not as active as they had been in the past. The California Teachers Association, for example, had spent far more money in past campaigns, while the prison guards, who benefited more than any other union under Davis, pretty much sat out the recall election. Other law enforcement groups also hung back. The diffidence of some unions may have reflected their discontent with the cautious Davis—or a sense that he'd lose and the money would be wasted. Others, sometimes reluctantly, gave more, not out of great expectations of a Davis victory but because they knew that even if he lost, during his last weeks in office he'd be signing or vetoing bills which they had sponsored. Pay-to-play was still in effect.

Senate Bill 2 is an example. One of several bills to extend health care coverage to more uninsured Californians, SB 2 was sponsored by John Burton, the powerful president pro tem of the senate and a close ally of organized labor, for which health care was a top priority. To the chagrin of

California business leaders, SB 2 put much of the burden of health care on employers. The bill was passed by both houses of the legislature and sent to Davis in September, just as the recall campaign intensified. Would he sign it, even though business opposed it and many saw the legislation as well intentioned but flawed? Or would he veto it and offend labor, his strongest supporter in his hour of greatest need? (To some extent that support was calculated to get his signature on SB 2.) With health care regularly listed by public opinion polls as a top voter concern, could approval of the legislation turn back the recall tide? Davis signed SB2 just days before the election; it wasn't enough to save him.

Business: A Winner at Last

Business interests had mixed success with Gray Davis. Corporations and business organizations were a major source of campaign funds for both of his campaigns for governor, with major contributions from corporations ranging from Enron, Chevron, PG&E, and Southern California Edison to AGI Management Corporation ($300,000 in 1998). For a Democrat, Davis had not been a bad governor for business. He showed interest in and support for the high-tech and entertainment industries, supported tax breaks for some businesses, and promoted California businesses abroad. It could even be argued that his use of the veto kept the liberal Democrats in the legislature in check on some business-related issues. Many of his liberal colleagues thought of Davis as a "business" Democrat.

In the good times, with Democrats dominating state government, business reluctantly went along. But when the recession followed the energy crisis and the budget deficit hit, business began to express its discontent. California was no longer "business-friendly," they claimed. High taxes, a high cost of living, declining services, and government regulations discouraged new business from coming to California, they asserted, and, worse, drove existing businesses out of the state and even out of the country. The evidence that this was happening is irrefutable, yet new businesses continue to form and thrive in California as well as in the eighteen states that impose higher tax burdens on individuals and businesses.[6] But taxes weren't the only issue of concern to California business leaders. Rising energy costs—blamed partly on the governor—remained a problem. The high cost of workers' compensation insurance in California was a painful thorn in the side of business. A new law requiring paid family leave for workers also added to business costs. And state-mandated health insurance was looming.

Business groups like the California Chamber of Commerce and the California Business Roundtable were outspoken on these issues and increasingly frustrated by the direction state government seemed to be taking. Republican leaders and the editorial pages of some of the state's newspapers were soon repeating the mantra "California is not business-friendly." Although some business leaders conceded that the budget Gray Davis proposed, including some tax increases, was superior to the compromise budget he ultimately signed, others shored up Republican legislators' resistance to tax increases as part of the solution to the budget deficit.

Most business leaders just want to get on with making money. But when government began to seem to hinder rather than further that goal, more and more business leaders turned hostile. But what could they do to bring change? The recall seemed improbable. Bill Simon and Darrell Issa didn't appear to be candidates who could win or accomplish much if they did. Former mayor of Los Angeles Dick Riordan was a businessman himself and so one of their own, but he had been a disastrous candidate in 2002 and seemed diffident about running again. All that changed on August 6, when Arnold Schwarzenegger announced that he would be a candidate.

Why would business groups line up behind a bodybuilder and movie star for governor? At first they didn't. They waited and watched to see what would happen, and gradually most were won over. Arnold Schwarzenegger was not unknown to the business leaders of the state—not only because he was a movie star but also because he had been moving in business circles for some time. Schwarzenegger, who is active in the management of his own extensive investments, has a reputation as a shrewd businessman. He made a point of getting to know many business leaders during the 2002 campaign for the initiative he sponsored for after-school programs. He also became acquainted with conservative intellectuals at Stanford's Hoover Institution and with the Republican leadership of the state, including former governor Pete Wilson.

Very quickly after announcing his candidacy, Schwarzenegger put together a team of economic advisers, including financier Warren Buffett, former Secretary of the Treasury George Shultz, Hoover Institution economists, and others. He named Pete Wilson co-chair of his campaign. Just by dropping a few names he had sent a message to business. He followed with talk about high taxes and excessive regulation, targeting workers' comp. His solution to the recession and budget crisis was to lighten taxes and regulations on businesses so that they could create jobs that would generate taxes. While this was a long-term solution to an immediate and urgent problem, business leaders and many voters responded positively.

Moments after he announced his candidacy, Schwarzenegger made a surprise phone call to a meeting of moderate Republican business leaders in a group called the New Majority. As journalist Daniel Weintraub describes it, "The call was unexpected, and a speakerphone was brought into the room so

Table 8.1

Top Ten Donor Groups for Arnold Schwarzenegger

1. Real estate, development, construction	$2,470,371
2. Financial services industry	1,589,961
3. Transportation equipment/sales	1,007,733
4. Communications/technology	724,414
5. Agriculture/food distribution	667,482
6. Lawyers	327,290
7. Film/entertainment/sports industries	323,975
8. Health care industry	272,200
9. Doctors and dentists	173,719
10. Lodging industry	129,250

Source: California Common Cause; California Secretary of State.

Schwarzenegger could address them all at once."[7] A week later they endorsed him—a significant start on the road to business support. Later the California Manufacturers and Technology Association and the California Chamber of Commerce endorsed Schwarzenegger. The latter endorsement was particularly notable. Deeply involved in state politics and public policy, the statewide chamber focuses mostly on lobbying. This was its first endorsement of a candidate for statewide office. As with labor, some of the clout of the chamber comes from its local chapters, although they cannot marshal the volunteers that labor does.

Business support gave Schwarzenegger's simplistic analysis of California's economy credibility. Business support also provided campaign funds. "I will go to Sacramento and I will clean house," Schwarzenegger promised when he announced his candidacy. "I don't have to take money from anybody. I have plenty of money myself."[8] He used around $8.6 million of his own money, but he also took a lot of money from business and other supporters, all the while campaigning against "special interests" (see Table 8.1). "Special interests have a stranglehold on Sacramento," Schwarzenegger declared in a campaign ad. "Here's how it works: Money comes in—favors go out. The people lose. We need to send a message: Game over."

According to campaign finance reports, Schwarzenegger's top donors were from the real estate, development, and construction industries. Stockton builder A.G. Spanos gave $344,700. A financial business that contributed to him issued a statement saying "American Sterling Corp. supports individuals who can bring fiscal responsibility, accountability and change to what we believe is a very serious crisis in California."[9] The Schwarzenegger campaign conveniently distinguished between its financial supporters—interested only in "fiscal responsibility and accountability"—and "the special interests of Sacramento,"[10] particularly labor and tribal casinos.

105

Of course those supported by labor and others don't see their allies as special interests either. But the voters often do, and the role of special interests and pay-to-play politics in Sacramento was a concern in the recall election. Schwarzenegger appeared to be the candidate least tied to traditional special interests.

Gray Davis knew he was in trouble with business and readily signed a bill reforming workers' compensation insurance—on September 30, 2003. It was far too little, far too late for business, especially with SB 2 still hanging over them.

When Schwarzenegger won, his affiliation with business interests was apparent in his every move. His transition team was diverse, but business leaders were prominent. On his first visit to the capitol after the election, he met with the transition team in the offices of the California Chamber of Commerce. He appointed their lead lobbyist as his legislative secretary. "Rather than cleaning house," a critic said, "this is far more like turning over the keys to special interests."[11] Schwarzenegger scheduled a Sacramento fund-raiser 2 weeks after his inauguration.

Indians: California's Newest Players

One set of "special interests" seems to have almost played into Schwarzenegger's hands, probably scuttling the candidacy of Cruz Bustamante in the process. In the 1990s, Indian tribes began establishing casinos on tribal lands, which, as sovereign states, are not subject to state taxes or regulation. But by state and federal law, the tribes must negotiate "compacts" setting the conditions for the operation of their casinos. California's tribal leaders learned quickly that good relations with governors are important, and as the money from their casinos rolled in, they began investing in making political friends.

Tribal gaming has become a $5-billion-a-year industry—and by 2003, the tribes had become the biggest single source of campaign funds in California. Seven tribes spent $13 million to prevent Arnold Schwarzenegger, who had promised to get more of their gambling revenues for the state, from becoming governor. Millions went to Davis's "No on Recall" campaign. The tribes also gave to conservative Republican Tom McClintock, an old friend and ally, they said, although cynics thought they were only helping McClintock in hopes that he would siphon votes away from Schwarzenegger.

But Cruz Bustamante was by far the biggest beneficiary of Indian gaming money—an ostensible advantage that turned out to be the kiss of death. Long an ally, friend, and defender of California's Native Americans, Bustamante has been a consistent and loyal advocate—even when they were poor, he says. One tribe gave Bustamante $2 million, another $4 million. Some of the funds went to the official Bustamante campaign; others went to independent campaigns on his behalf. And in an effort to get around campaign contribution rules, some went to Bustamante's 2002 lieutenant governor campaign committee. The move backfired when a judge ruled that the use of these funds violated campaign finance laws. Bustamante shifted the money to the "No on 54" campaign, which, in turn, used the money to pay for television ads featuring . . . Cruz Bustamante!

The media reported the donations, Schwarzenegger and Arianna Huffington attacked, Bustamante defended himself weakly, and voters noted the hypocrisy of the Proposition 54 ploy. The lieutenant governor and his campaign manager may have hoped voters wouldn't notice or care about their special connection to this special interest, but the voters did and it soured many on Bustamante as an alternative candidate. Aside from his own lackluster performance, it was surely the major factor in his defeat.

Follow the Money

Indian tribes, labor, and business were the top three contributor categories in the recall election, topping $10 million each. The tribes and labor gave almost exclusively to the "No on Recall" and Bustamante campaigns, while business gave almost exclusively to Schwarzenegger.

But they weren't the only donors. Trial lawyers are another group that invests heavily in most elections and the recall was no exception, with hundreds of thousands of dollars going to "No on Recall" and Bustamante, because Republicans are usually not supportive of trial lawyers' interests. As Table 8.1 indicates, however, lawyers were one of the top ten sources of funds for Arnold Schwarzenegger, too.

Other groups that usually contribute abstained or gave minimally in this election. Prison guards and other law enforcement groups as well as doctors and other professional groups along with some business groups "sat on their wallets." Jim Knox of California Common Cause says they may have been put off by new campaign contribution laws, or they may have thought Schwarzenegger didn't need the money. Maybe they didn't want to be labeled "special interests." Or perhaps they were just uncertain. "A lot of the traditional interest groups in the Capitol tend to play it safe when it comes to contributions," said Kim Alexander of the California Voter Foundation. "They weren't sure what strategy would be effective" in this unique situation.[12]

One source of uncertainty was the new law limiting campaign contributions to individual candidates to $21,200. Approved by the voters in 2000, the recall was the first election to test Proposition 34. Like most attempts to limit campaign spending, it had mixed effects, as candidates and donors found

Table 8.2

Top Candidates and Committees in Campaign Spending

Arnold Schwarzenegger	$19,712,635
Californians Against the Costly Recall	17,426,717
Cruz M. Bustamante	14,446,725
Peter Ueberroth	3,944,778
Arnold Schwarzenegger's Total Recall Committee	3,467,558
Rescue California . . . Recall Gray Davis	3,058,046
Taxpayers Against the Governor's Recall	2,871,557
Tom McClintock	2,343,617
Darrell Issa	1,586,956

Source: California Common Cause, www.recallmoneywatch.com.

ways to get around the law. Candidate campaigns followed the letter of the law, but the committees defending Gray Davis were not subject to the limits and other independent committees formed to support other candidates. As a consequence, figuring out who gave how much to whom is a challenge. All the records are available through the secretary of state (www.ss.ca.gov), but California Common Cause, a reform group, has done the best job of trying to analyze and present the data in an understandable way. They report total expenditures of $83.6 million dollars—just a little less than what was spent on the gubernatorial election of 2002. That's a disappointing and somewhat surprising total. People hoped that the shortness of the campaign, the heavy focus of the news (and other) media, and limits on campaign contributions might lower total expenditures. They didn't, perhaps partly because there were so many candidates in the field. Table 8.2 reports the totals raised by the top candidates and committees. Arnold Schwarzenegger himself was the biggest single donor, at $8.6 million, but at least three other candidates (Peter Ueberroth, Darrell Issa, and Garrett Gruener) gave a million dollars or more to their campaigns.

Personality Politics: Movie Stars and Political Stars

Politics is about individuals as well as groups, and several stars played parts in the recall campaign of 2003. Arnold Schwarzenegger himself took center stage and held it throughout the campaign. A few other Hollywood stars, such as Rob Lowe, eagerly supported him, but others held back or even supported one of his opponents. Danny Glover, for example, campaigned with Gray Davis in the final days, and Arianna Huffington had a lot of friends in Hollywood. Oprah Winfrey and Larry King played their parts, too, by giving Arnold Schwarzenegger uncontested opportunities to frame the campaign on his own terms. All this added a touch of glamour—as well as media attention—that's usually lacking in gubernatorial elections even in California.

The political stars were all out, too. Democrats lined up solidly if somewhat reluctantly behind Governor Davis, guided by U.S. Senator Dianne Feinstein, the senior and most popular of the state's Democratic leaders. Feinstein opposed the recall from the outset, refused to become a save-the-party candidate herself, and performed in several television ads against the recall. Other leading Democrats spoke out against the recall but held back a bit in the campaign. California's Democratic congressional delegation took the lead in encouraging Cruz Bustamante to run as an alternative candidate, then appeared to do little to assist him during the campaign. Bill Clinton, Al Gore, and several of the candidates for the Democratic nomination for president in 2004 toured with Davis, but none of the Clinton magic rubbed off on him and the others had little magic to spare. In the last days of the campaign, Attorney General Bill Lockyer, Superintendent of Public Instruction Jack O'Connell, and other California Democratic leaders joined Governor Davis for rallies around the state—mostly attended by union activists and volunteers.

The Republican establishment—small and lacking in stars as it is—eventually rallied round Schwarzenegger. With several Republicans running at the outset and McClintock in the race to the finish, some waited until Schwarzenegger's candidacy took off. Support from Republicans in the California congressional delegation grew as the campaign proceeded. Others were there from the outset, including former governor Pete Wilson and key strategists from his administration, such as Sean Walsh, Bob White, and Rob Stutzman. Wilson proved both an advantage and a modest liability to Schwarzenegger, bringing instant support and credibility with the Republican and business establishments but also fear and loathing from labor and minority voters. He remained co-chair of Schwarzenegger's campaign throughout (along with Latino legislator Abel Maldonado) but quickly lowered his profile after the campaign launch, purportedly at the suggestion of Maria Shriver—another star associated with the Schwarzenegger campaign.

A Different Sort of Election

Interest groups have traditionally been important in California politics. They shape policy, they support candidates, they rally their members at election time, and they fund campaigns. But the "Great Recall of 2003" was a different sort of election. A little less money was spent and some groups kept their checkbooks closed out of uncertainty, caution, or both. Money from individual donors was almost as important as money from groups.

In the end, as exit polls showed, most groups appear to have had less influence over how their members voted than in the past, with substantial numbers of women, Latinos, and union members voting for Schwarzenegger. Money mattered less and so did group endorsements and campaigns. Voters appear to have made their minds up early and independently. Perhaps it was a voter revolt not only against the incumbent governor and politics as usual but also against the traditional roles of interest groups, money, and political leaders in elections.

9

The Campaigns and the Media

Whose Election Is It Anyway?

The media are important in any election; they perform the vital role of connecting the voters, issues, and candidates. In a state as large as California—with 35 million people, 15 million registered voters, and over a dozen distinct media markets—media linkages are crucial to providing much needed information. In the early days of the recall effort, nontraditional media such as talk radio and the Internet promoted the petitions. As the campaign unfolded, traditional media outlets such as newspapers and television became major conduits of information. By campaign's end, Californians and the world had witnessed coverage by three almost separate branches of the media: the traditional campaign media, the traditional news media, and the entertainment news media, or what some have called the "celebritization of politics" resulting from the presence of a movie star candidate.

The election and campaign of 2003 differed from others because of these distinct media. The difference was the product of unique circumstances. The recall of a sitting governor, a rare event in itself, brought extra attention. And this time it was California, not North Dakota—the only other case, which had occurred 80 years earlier. The brevity of the campaign was another difference. Most high-profile American election campaigns sprawl out over a period of 18 months to 2 years. This one lasted just 75 days, a compression of time rarely experienced in the American political system, though not so uncommon in other countries. The final distinction was the presence of a celebrity candidate of unusual star power.

The combination of these three—a rare recall, a short campaign, and a movie star candidate—made for a most unusual election, drawing as much voter interest and media coverage in a state election as is normally seen in a presidential election. Maybe more. But beneath these unique elements were also the elements of traditional campaigns.

Added to all this was a series of cliffhangers that left Californians and the media breathless for months. Week by week, even day by day, events occurred that kept us in a state of uncertainty and anticipation in an election for which there was no precedent. Would the recall qualify? Would Arnold Schwarzenegger run? Would a major Democrat join the list of replacement candidates? Would court cases delay the election or alter its form? Would Gray Davis resign? Would Republican candidates other than Schwarzenegger withdraw to improve the chances of unseating Davis? Would Schwarzenegger debate? When he did, would he self-destruct or prove himself? Would the last-minute stories of Schwarzenegger's misbehavior with women seriously damage his candidacy? And of course the ultimate cliffhanger, could the movie star govern once elected?

All this added up to a roller coaster of a campaign, with much uncertainty, lots to speculate about, and a great deal of interest. No wonder few elections in history have drawn such media attention—or aroused such controversy.

The Campaigns: Shaping the Messages

While the media spend much of their energy reporting on events, *what* they report is often shaped by campaign consultants and *their* messages. In some elections, traditional techniques mostly involve "retail" politics—direct, personal contact between candidates and voters, or at least between volunteers for candidates and voters. These include door-to-door canvassing, neighborhood coffees, phone calls to voters, and "town meetings." California is too big for much of this kind of campaigning, and this election cycle was too short for the campaigns to organize much. Most town meetings were put on as shows for the media and actually reached few voters personally. Only labor unions in some parts of the state—especially Los Angeles, San Jose, and San Diego—made much effort to make personal contact with their members and supporters, in their case on behalf of Gray Davis and Cruz Bustamante. Unions, in conjunction with the Democratic Party, did organize get-out-the-vote (GOTV) drives for supporters (as identified by phone banking), and the Republican Party and some conservative groups also made some effort at GOTV, but that was about the sum of person-to-person, retail politics in this campaign.

Mass campaign media—wholesale politics—were far more important in the recall election than they typically are in statewide campaigns in California. "Wholesale" techniques are a cheaper, albeit less personal way to reach large numbers of voters. The primary means to do so are through direct mail and television advertising, although recently campaigns have added automated phone messages from big-name supporters. Toward Election Day, for example, many Democratic voters received messages from Dianne Feinstein,

Barbara Streisand, and Bill Clinton. Perhaps as many were irritated by these calls as were pleased. Radio advertising is common, too. It's cheap and easy to target to particular constituencies such as Latinos on Spanish-language stations or fans of particular talk radio hosts. When Bustamante was criticized for accepting so much money from tribal gambling interests, Schwarzenegger was heard in radio ads saying, in his own unique voice, "All the major candidates take their money and pander to them. I don't play that game."

But these are minor methods compared to direct mail and television. Direct mail campaigns are popular because they can be targeted to particular voters with particular interests—just Democrats, Republicans, Latinos, or union households, for example. Both the California Labor Federation and the Schwarzenegger campaign sent mailers to remind presumed supporters to vote and, because of the unusual nature of the recall ballot, with 135 candidates, both mailers reproduced the entire ballot, calling attention to their candidates. The labor mailer went to 450,000 union households where phone banks had identified undecided voters and was printed in 115 versions because by state law the order in which candidates were listed varied from district to district.[1] Other mailers, mainly opposing the recall or advocating for Schwarzenegger and Bustamante, made a straightforward case for their candidates. Nor did the campaigns overlook absentee voters. The Schwarzenegger campaign and others sent early mailings to encourage these voters to get their ballots in, perhaps a particularly productive tactic for Schwarzenegger, given the news about his past bad behavior that made headlines in the last days of the campaign. In fact, by the campaign's end, nearly 30 percent of all votes came via absentee ballot.

But in California, television advertising trumps all other campaign techniques, and the recall election was no exception. Roughly half of the $84 million spent by the various campaigns went for television advertising, with $10 million spent in the last week. The Schwarzenegger and Davis campaigns led in spending on television, followed by Bustamante's. Ads for other candidates rarely flickered across our screens. A sampling of the ads:

• U.S. Senator Dianne Feinstein (opposing the recall): In one of the two ads she did, Feinstein says, "This governor was elected just last November. The recall was started within three months by people who were unhappy with the outcome of an election in which 8 million Californians voted. The recall is creating uncertainty and instability—it's bad for our economy, it's bad for jobs and it's bad for California. I'm going to vote against it—and I hope you will, too." She does not mention Gray Davis.

• Arnold Schwarzenegger (for himself): In two commercials with similar themes, Schwarzenegger promises "to change things in Sacramento," promising to "audit everything, open the books, and then we end the crazy deficit

113

spending. . . . It comes down to this. If you are happy with the way things are, then keep your current leaders. If you want to change this state, then join me."

• Gray Davis: "Newspapers are calling it a circus. Millionaires, local gadflies, political mavericks, even a porn king—all running for governor. . . . Imagine the possibilities. Where a replacement might stand on the issues. Their qualifications. How the uncertainty would impact our economy, even make things worse. A lot's at stake. The future of the world's fifth largest economy."

• Cruz Bustamante: "Arnold doesn't share our values. He won't fight for our health care, our neighborhoods, our jobs. He doesn't live in our world. He lives on Planet Hollywood. There is a long list of candidates. Finding my name won't be easy. But I need you to do it—because I need your vote for governor."

• Gray Davis: "Have questions about Arnold Schwarzenegger? So do a lot of people. He ducks tough questions. Didn't vote in 13 of the last 21 elections. And now he refuses to debate the governor he's trying to replace. Vote 'no' on the recall."

In other ads, Schwarzenegger referred to his own immigrant background and the success he found in California, appealing to Latino and Asian voters with similar experiences. He also emphasized education and programs for kids, which he had advocated in the past. In still other ads, he talked about the "tremendous disconnect between the people of California and the leaders of California." In all these ads, Schwarzenegger spoke directly and comfortably, albeit in his monotonal accent, to the camera—and to the voters. Other ads featured former New York City Mayor Rudy Giuliani endorsing Schwarzenegger or off-screen announcers telling his story.

Gray Davis appeared less frequently in ads opposed to the recall, presumably because his personality was part of his problem. His most prominent ads featured Feinstein; in others an off-screen announcer spoke the words. When he did appear, he demonstrated no growth in media skills.

With less money, Bustamante was a lesser presence on television. In his own ads, he spoke of his "tough love" plan for California—spending cuts and tax increases. He actually appeared more frequently in ads against Proposition 54, the "racial privacy" initiative, usually speaking somewhat stiffly to an apparently rapt audience. These ads were due to his shift, to avoid legal difficulties, of tribal gambling funds to the "No on 54" campaign. The ads may have backfired by reminding some voters of this shifting of funds and others that Bustamante was the candidate of liberals and minorities.

As election day approached, the ads got nastier. An ad funded by the Republican Governors' Association showed the faces of Gray Davis and Cruz Bustamante morphing into one. Davis struck back with an ad quoting the

114

San Francisco Chronicle on Schwarzenegger's lack of experience and avoidance of questions from the press. Overall, however, observers judged the campaign ads as "relatively tame." "This is not exactly the politics of personal destruction," said one communications expert. "The stuff we're seeing is factually based as opposed to 'The other guy's scum.'"[2]

Most of this advertising was concentrated on Southern California, where Los Angeles television stations reach up to 44 percent of the voters. Viewers in the San Francisco Bay Area saw far fewer ads (perhaps contributing to lower turnout there) and at times may have wondered where the campaigns were. September campaign spending reports showed Schwarzenegger spending only 16 percent of his television advertising budget in the Bay Area, with Bustamante at 22 percent and Davis at 28 percent. Besides the scale of the Southern California media market, the Bay Area is heavily Democratic, leading Democrats to take it for granted and Republicans to view advertising there as a poor investment. "Every candidate has to make a decision to hunt where their ducks are, and for Republican candidates, those ducks are not typically in the Bay Area," Republican campaign consultant Ray McNally told the *San Jose Mercury News*.[3]

The television advertising, direct mail, and every other element of the campaigns were controlled by the overall message each campaign sought to convey. For Gray Davis, the message to voters was to play it safe with an experienced candidate rather than risk change on an unknown and further disruption of the state's economy and politics fomented by a Republican conspiracy to overthrow his election. He attempted to soften his image while at the same time conveying the image of a hard-working governor tending to the people's business by signing legislation, albeit often catering to key Democratic constituencies. Both Bustamante and Davis sought to shore up the Democratic base, which—with 1.3 million more registered Democrats than Republicans—could have been enough to win the election had they succeeded. Bustamante presented himself as the safe Democratic alternative, appealing especially to Latinos and liberals. Tom McClintock presented himself as the pure Conservative, rallying that base while generally avoiding a frontal assault on his main Republican opponent. Schwarzenegger played the outsider and candidate of change, successfully presenting himself as a moderate as well as the Republican alternative most likely to bring victory. All four major candidates attempted, sometimes with limited success, to stay on message in their paid advertising and interaction with the news media.

That didn't keep the campaigns from feeding the media information calculated to damage their opponents. The Davis campaign, while trying not to confirm its candidate's reputation for practicing puke politics, passed on information harmful to Darrel Issa and Arnold Schwarzenegger, while

Schwarzenegger frequently reminded voters of Davis's past acts of negative campaigning, and a consultant associated with the Schwarzenegger campaign distributed information calculated to harm Bustamante.[4]

The Polls: What Did We Know and When Did We Know It?

Another key element of every campaign and election is the public opinion poll. Candidates commonly have their own internal public opinion polls to see which ideas appeal to voters and which do not. Based on the data produced by the polls, candidates and their consultants shape campaign strategies. As with everything else about the 2003 recall campaign, the polls seemed to be on a roller-coaster ride. It was sometimes difficult to tell which strategies would be most successful.

Besides polls, campaigns relied upon focus groups—small groups of typical voters who are guided through a conversation about candidates and issues by professional consultants seeking to assess voter reactions to campaign themes, issues, images, and the wording of campaign slogans and ads. Schwarzenegger campaign focus groups, for example, responded well to Gray Davis's approach to public education but reacted negatively to the governor's claims about "the right-wing effort to take over the government." Schwarzenegger scored with his tough statements on politicians who sell out to special interests and on Californians' tax burden.[5]

These ideas shaped the messages, with the campaign consultants following their impacts closely through daily "tracking" polls. Assuming the polls are valid, which isn't always the case, this means the campaigns and candidates know far more about the progress of the campaigns than the rest of us, including the news media. In June—as one of his consultants told a post-election forum—2 months before Schwarzenegger announced, his campaign poll had predicted victory if he "ran up the middle." Schwarzenegger's tracking polls also "showed him ahead all the way" in the replacement race throughout the campaign, despite the close race reported by outside polls.[6] Similarly, Davis's polls showed him losing, although there were times when the gap narrowed hopefully, including the last weekend, when the *Los Angeles Times* ran its articles on Schwarzenegger's alleged groping of women.

External polls—surveys conducted by organizations independent of the campaigns—provided conflicting results throughout the campaign, thus adding to the confusion. Among them were The Field Poll, California's oldest and most respected poll, the *Los Angeles Times* poll, and the Public Policy Institute of California (PPIC) Statewide Survey.[7] CNN, USA Today, and Gallup also polled, as did the California Chamber of Commerce and other interest groups. One poll showed the recall winning by 64 percent at one point; another, just days later, showed the "Yes on Recall" vote at 50 percent. One spelled doom for Davis, the other gave him a fighting chance. Early on, most polls put Bustamante in the lead among the replacement candidates, but some put Schwarzenegger there. Opposition to the recall and support for Bustamante were generally strongest in the *Los Angeles Times* poll, while the Field Poll showed more enthusiasm for the recall and a closer race between Bustamante and Schwarzenegger. As the campaign came to a close, however, all the polls pointed to victory for Schwarzenegger and defeat for Davis, though none predicted the magnitude of Schwarzenegger's victory.

Ultimately, the polls may have reflected a volatile electorate and voter confusion. Or perhaps they contributed to the confusion. John Feliz, a campaign strategist for Tom McClintock, says the *Los Angeles Times*'s polling was "way off all the time."[8] Strategists for the other campaigns seemed to agree, sure that their own polling was better. What was the problem? Polling in California has become more difficult because it's hard to find Californians at home, answering rather than screening calls and willing to talk to a pollster for 20 minutes in a language both parties understand. It has also become harder for pollsters to find the people who will actually vote rather than just saying they will.[9] But the ultimate problem for pollsters was the complexity and uniqueness of the recall election. Voters were unfamiliar with the process and many of the candidates. Their confusion may have affected polling and their opinions may have fluctuated more than in a normal election, when party identification is the strongest predictor of voter behavior. The polls were right all along about voter dislike of Gray Davis, which skewed traditional party loyalties, as did the star power of Arnold Schwarzenegger and the electorate's general frustration with "politics as usual." In the end, despite their inconsistencies, both the internal and external polls predicted the outcome if not its proportions.

The Debates

Debates have become a fixture in American elections, despite doubts about their impact or significance. In a September 2003 PPIC survey, 67 percent of the respondents said that debates were somewhat or very important to them in deciding how to vote in the recall election.[10] But the real forces behind debates are usually the news media, good government groups like the League of Women Voters, and candidates lagging in the polls and looking for free air time. Audiences are rarely large and few debates appear to have changed the minds of enough voters to make a difference in the outcome of elections. More often, voters seem to find evidence in debates to confirm their previous candidate preferences—an outcome that may be far from insignificant. That may have been exactly what happened in the recall election.

Several debates were held over the course of the campaign, mostly featuring Bustamante, Camejo, Huffington, and McClintock. These got scant attention beyond local media, but two others were more widely watched and more significant.

The first took place in Walnut Creek near San Francisco and topped the ratings in the Bay Area and Los Angeles, with excerpts shown on television news programs around the state. Bustamante, Camejo, Huffington, McClintock, and Peter Ueberroth debated with civility and good humor, each demonstrating his or her unique strengths. Both the media and some of the candidates noted the absence of Arnold Schwarzenegger. But what made this event unique was the presence of Gray Davis doing Q&A with a journalist before the candidates debated. Attempting with some success to project a warmer personality, Davis defended his record but also recognized shortcomings and promised to do better at staying in touch with the public. But the only major outcome of the first debate was a poor performance by Ueberroth, leading to his withdrawal from the race a few days later. Two other debates followed, without Davis or Schwarzenegger. At each, Bustamante grew more aggressive in his attacks on the absent Schwarzenegger. "This 'trust me' politics doesn't fly. It doesn't work," he said. "I just wish we could get Arnold here."[11]

Meanwhile, Schwarzenegger's handlers kept him out of the debates and away from the press as much as possible, partly because he could get coverage without such direct contact and partly to help him get up to speed on the issues. A consultant to Schwarzenegger said taking part in more debates would have weakened his outsider image and made him and his campaign look "conventional."[12] Dismissing the first debates as "warmups,"[13] Schwarzenegger agreed to participate in one debate on September 24, less than 2 weeks before the election. The press and a curious public immediately focused on that event. Would Schwarzenegger embarrass himself? Would he vanquish his foes? How would the others stand up to the Terminator, by then leading in the polls?

In a widely watched, freewheeling debate, Schwarzenegger did survive and comported himself reasonably well. As with George Bush in the debates with Al Gore, expectations were so low that Schwarzenegger had little difficulty meeting or exceeding them. Well-rehearsed but seemingly at ease, he neither misspoke nor embarrassed himself. "Debates are job interviews," said an academic expert, "and you could see him in this job based on the performance he gave."[14] More entertaining than informative, the debate featured a spirited exchange of insults between Schwarzenegger and Huffington, articulate advocacy of conservative and liberal policies by McClintock and Camejo, and a dull performance by Bustamante, who failed to engage either his opponents or the viewers.

Thanks in part to Schwarzenegger's presence, the debate drew huge media coverage and a large audience. Five hundred journalists reported on the event, which was broadcast live throughout California and repeated on cable networks as well as in England and Japan. The television audience was double normal for the time slot in some areas. "The debate . . . actually drew more people than the premier of *West Wing*," said communications professor Barbara O'Connor of California State University, Sacramento. "When was the last time you could say that about a political event in California?"[15]

Although internal campaign surveys showed Tom McClintock, the articulate conservative, winning every debate, campaign consultants across the political spectrum agree that the September 24 debate "turned the tide" for Schwarzenegger (although it was already running his way).[16] Schwarzenegger surged in subsequent polls as Bustamante, his main replacement competitor, faded—just as he had in the debate.

The News Media: Kill the Messenger

Tension between the news media and candidate campaigns is inevitable in elections. Each seeks to control the other. Candidates and their consultants seek to manipulate the media to their advantage, with staged events and canned comments. Reporters try to cut through that control to get at the real candidate and what they view as the real issues—often seen as avoided by the candidates. The news media often win this struggle, since they filter the news to the public; the only recourse for candidates is paid advertising, which many voters view with skepticism.

But in this election, the news media had less control than ever, and at the same time coverage of the campaigns was more extensive than ever. Two factors contributed to this. The brevity of the campaign gave editors and reporters less time to prepare, plan their coverage, and generate stories, and they were driven by the ever-changing schedules of the candidates and by events beyond their control. In a short time, media managers had to reallocate significant resources for the recall campaign, often including reporters who were ill prepared and had no time to be brought up to speed. But an even more important factor in shaping news coverage of the recall election was the presence of Arnold Schwarzenegger as a candidate—and the media strategy adopted by his campaign.

"There was a conscious decision that we were going to play to the people, not the press," said Schwarzenegger media consultant Don Sipple. "It's well known that in most campaigns, the press wags the dog. [This] was in keeping with the unconventional nature of this campaign."[17] Schwarzenegger's consultants admit that they and the candidate were not ready at first, in

119

terms of both briefing the candidate and setting up their media infrastructure. They made mistakes early on, letting Warren Buffett criticize California's sacrosanct Proposition 13 and letting the candidate struggle to answer questions in his first television interviews. But they soon took control, holding Schwarzenegger back from journalists while scheduling regular appearances and events journalists couldn't avoid covering—all calculated to maximize television coverage. Schwarzenegger waved a broom in front of the state capitol to accentuate his message about cleaning out politics as usual. He stood by as a car was smashed to illustrate what he planned to do to the hated car tax. He got media coverage for blatantly staged "town hall meetings." Ordinary candidates, lacking Schwarzenegger's star power, couldn't get away with this. When they staged an event or called a press conference, they were lucky to get a few reporters and a camera or two. When Schwarzenegger did, cameras and reporters from all over the world filled the room, often sidelining veteran political reporters who might have asked tough questions if they'd had a chance.

At the same time, Schwarzenegger gave few interviews to political reporters and traditional news organizations. "A lot of thought was given to the idea that we had the ability to expand upon the traditional outlets that tend to cover politics and take our message directly to a segment of the population that frankly isn't the 'Meet the Press' crowd," explained Schwarzenegger media consultant Todd Harris.[18] No interviews with mainstream political journalists were scheduled during the first weeks of the campaign and many interview requests were declined. Unlike other candidates, Schwarzenegger didn't even bother to meet with the editorial boards of some major newspapers, apparently unconcerned about their endorsements. His media spokesman pointed out that some of these newspapers had run critical stories while also denying that this was the reason interviews were declined.[19]

Meanwhile, the Schwarzenegger campaign staged major, highly visual events calculated to draw coverage and send its own messages to the voters. Early in the campaign, they convened an "economic summit" with noted academics and respected elder statesmen such as Buffett and Shultz. Little came of the summit in terms of specifics, but the event suggested Schwarzenegger's serious consideration of the voters' top concern and helped legitimize his candidacy. "My Economics," an essay by the candidate, followed in the *Wall Street Journal*.[20] Again, specifics were few but the message was clear. As Republican consultant Dan Schnur observed, "Schwarzenegger made people comfortable with the idea that he could govern. The lesson is that substance matters, or at least the appearance of substance."[21] A few carefully staged town hall meetings gave the impression that Schwarzenegger was meeting with masses of voters. Visuals of Arnold being mobbed wherever he went conveyed his popularity. Toward the end of the campaign—as polls showed him taking the lead—he announced plans for his first hundred days in office, casting "an aura of inevitability around his candidacy."[22]

Finally, using a more traditional campaign tactic, he took his show on the road in a bus tour of California, with more mob scenes everywhere he went and plenty of television coverage. The structure of the tour tells a lot about his candidacy. The candidate's bus was called *The Running Man*, after one of his movies. His followers were in a bus called *Total Recall*. Around two hundred reporters and photographers from all over the world packed into four more buses named *Predator 1* through *Predator 4*, followed by still more journalists in cars and vans. The tour was a media event matched only by presidential campaigns in their final stages.

One result of all this was unprecedented news coverage. The recall campaign made the front page of every major newspaper in the state every day, with full-page features inside and frequent editorials. Despite the skepticism of newspaper journalists about the candidacy of movie star Schwarzenegger, a Stanford University study reported that even newspapers in the heavily Democratic Bay Area gave Schwarzenegger massive coverage, including 43 percent more headlines than Cruz Bustamante got in September and an astounding 523 percent more in the final 11 days of the campaign. Schwarzenegger's headlines exceeded those of Gray Davis by a much smaller margin, but throughout the campaign he scored far more photographs in Bay Area newspapers than any of his rivals.[23]

National news media—both print and electronic—followed the story. According to one report, ABC, CBS, and NBC gave six times as much coverage to the recall as to the Democratic candidates for president during August and September.[24] But most extraordinary, no doubt due to Schwarzenegger's presence, was the intensity of coverage by local television news. A study by the Norman Lear Center at the University of Southern California found that television stations in California's top media markets gave 0.45 percent of their newscasts (37 hours total) to the 1998 gubernatorial election. "My guess," said the director of the center, "is we exceeded that in the first few days of the recall."[25] The recall was news on almost every local newscast almost every day—perhaps more coverage than ever before in a California election. Satellite technology enabled local reporters to broadcast live from wherever events occurred, and "talking heads" (experts on politics) were in great demand—as we can assure you from personal experience.

The traditional news media covered the race in traditional ways, albeit scrambling to keep up with the campaigns and the TV crews. Reporters were quickly reassigned beats, public opinion polls were commissioned, television ads were dissected in "truth boxes," and editorial pages weighed in with their endorsements. Few major newspapers, however, endorsed the recall or Schwarzenegger. The *Los Angeles Times;* the *Oakland Tribune;* the Sacramento, Fresno, and Modesto *Bees*; the *San Francisco Chronicle;* and the *San Jose Mercury News* all recommended voting "no" on the recall. None of these except the *Oakland Tribune* recommended a replacement candidate, but they later changed their view. The *Long Beach Press-Telegram*, the *Orange County Register,* and the *San Diego Union-Tribune*—all in more conservative areas—recommended a "yes" vote on the recall and Schwarzenegger for governor.

The Schwarzenegger campaign's strategy to take its case directly to the people and avoid being "played" by the press worked through most of the campaign. Few news stories pushed Schwarzenegger to react, and only the articles in the *Los Angeles Times* about his groping record caused a stir. That news spread like wildfire around the state, the nation, and the world via every medium. Schwarzenegger responded with a quick apology of sorts and tried to proceed with his bus tour, although it was now tainted by constant questions about the *L.A. Times* stories. The *Oakland Tribune* withdrew its editorial recommendation of Schwarzenegger for governor and his opponents played the issue to the hilt. Schwarzenegger's wife, Maria Shriver, rallied to his side, however, and his supporters were unshaken.

The public seemed to turn on the *Times*, rather than on Schwarzenegger, in the great tradition of "killing the messenger" who brings bad news. Garry South, a strategist for Gray Davis, told fellow consultants that the Schwarzenegger campaign had prepared the public by warning of attacks to come, so people merely saw the news stories as more dirty tricks from Davis. The *Los Angeles Times* denied that the story was planted, insisting that they "published the story when it was ready," after appropriate research and confirmation had been completed, and that they had not delayed it to make it a last-minute "hit piece."[26] But many voters believed Schwarzenegger, not the highly respected *Times*, which was inundated by backlash e-mails and over a thousand canceled subscriptions. Campaign tracking polls showed that 94 percent of the voters had heard about the issue.[27] Schwarzenegger's popularity waned momentarily, but when Davis and others joined in the criticism, it surged again, just in time for Election Day.

All this added up to unprecedented media attention for the election and the candidates. For some without lots of money—especially Camejo and McClintock—that meant invaluable exposure they could not otherwise have hoped for. Both of these articulate candidates gained respect, and McClintock became something of a national hero to hard-core conservatives. The extraordinary media attention stimulated and probably also reflected voter interest. But while polls reported that over 90 percent of the voters were paying close attention to the race, the same polls also revealed that only a minority of voters—26 percent—relied on newspapers as their primary news source. While 46 percent cited television, 16 percent relied on radio and 8 percent got their news from the Internet. Advantage: Schwarzenegger. An impressive 83 percent said they saw television ads for candidates, but apparently they viewed the ads skeptically, because only 6 percent said they "were helpful in deciding how to vote."[28]

But in this election, traditional campaign techniques and the mainstream news media were not the only sources of information for voters. A third sort of media campaign was also under way.

Alternative Media: Celebrity Politics

"This beats watching *Friends* or a rerun on TV. I'll just watch California," said former president Bill Clinton. Lots of people felt like that—all over the world—and the media beyond traditional news outlets knew it and played to it.[29]

Talk radio—in California and nationally—was key to the recall from the outset to the conclusion. Tom McClintock and Arnold Schwarzenegger were frequent guests. Angry talk show hosts—or rabid dogs as some like to call themselves—kept their angry audiences engaged throughout and were particularly important in turning public opinion against the *Los Angeles Times* at the end, perhaps even saving Schwarzenegger. The Internet also continued to play a role as an alternative source of information, especially for those alienated from the mainstream media.

But the big difference in this campaign was the attraction of the movie star candidate for national and international media and for the entertainment media. Arnold Schwarzenegger appeared on all the national news programs; perhaps more important, he appeared on the *Tonight Show* with Jay Leno (announcing his candidacy) and with Oprah Winfrey, Larry King, and Howard Stern. All gave him major exposure and the appearance of media accessibility —without the tough questions that might have been asked by political journalists. The *Oprah Winfrey Show* was especially important because he was joined by his wife, Maria Shriver, who helped neutralize charges of sexist behavior. Other candidates got some national air time, but none came within minutes or even hours of Schwarzenegger.

Much of that time was on "entertainment" TV. Shows like *Entertainment Tonight*, *Access Hollywood*, and others fawned over the movie-star candidate and his glamorous wife, herself a television journalist. Reporters asked tough questions like "What's it like to vote for yourself for governor?" Opponents rarely got a mention. While these programs were almost totally free of overt political content, they gave viewers the important impression that they were getting to know Schwarzenegger as a person—more about his life, his wife, his kids, his Hummers, and even his measurements and his fashion choices. We didn't hear these things about Gray Davis or Cruz Bustamante, although we did get some fashion advice from Arianna Huffington (wear sensible shoes). "The power of the entertainment media eclipsed the serious media," observed Orville Schell, dean of the journalism school at U.C. Berkeley. "Nobody seemed to notice."[30]

Besides all this, the presence of the movie star and a huge number of virtually unknown, often eccentric other candidates in the campaign made it a joke to some people—which gave it even more publicity. Ninety of the fringe candidates appeared on Jay Leno's *Tonight Show*. Six were contestants on *Who Wants to Be Governor of California?* on the Game Show Network. David Letterman did more than one "Top 10" list relating to the recall on his *Late Show*. All of it was fodder for monologues for Leno, Jon Stewart (on the *Daily Show*), and Conan O'Brien. Here are, for example, Letterman's top 10 Schwarzenegger campaign promises[31]:

10. To do for politics what he did for acting.
 9. Combine the intelligence of George Bush with the sexual appetite of Clinton.
 8. A heaping tablespoon of Joe Weider's dynamic body shaper in every pot.
 7. Every freeway gets a dedicated car-chase lane.
 6. Seek advice from elder political statesmen like Jesse Ventura.
 5. Crack down on schools graduating students who can't bench press 180 pounds.
 4. Solemnly swear to support the constitution of Gold's Gym.
 3. Goofiest-named governor since Pataki.
 2. Raise the minimum age for dating Demi Moore.
 1. Speak directly to voters in clear, honest broken English.

Jokes aplenty, mostly about Schwarzenegger, but most were good-natured and it was all invaluable publicity, generally reinforcing his image as a nonpolitician and an action hero.

Clearly it was Schwarzenegger's celebrity status that brought this attention to the campaign.[32] No politician has ever been such a draw (although John F. Kennedy had his moments) and no other actor-politician has received the attention of megastar Schwarzenegger. But one medium that normally pays a lot of attention to movie stars took a pass on this particular celebrity candidate: the supermarket tabloids. Instead of appearing in the *Star*, the *Globe*, or the *National Enquirer*, headlines about Schwarzenegger's sexcapades were in the *Los Angeles Times*. The outrageously flamboyant tabloids kept quiet about Schwarzenegger, although the *Weekly World News* ran an exclusive story about extraterrestrial support for Schwarzenegger ("Alien Backs Arnold for Governor!"). Perhaps the tabloids were cautious because Schwarzenegger can be litigious, or because they viewed his constituency as their own. But another theory is that Schwarzenegger's business associate Joe Weider helped suppress negative tabloid coverage. Weider publishes muscle-and-fitness magazines for which Schwarzenegger wrote columns and recently made a "lucrative" business deal with American Media, which publishes two major tabloids. Allegedly, American Media promised Weider to "lay off" Schwarzenegger. Instead of attacking the body-builder candidate, American Media published *Arnold, the American Dream*, a glossy magazine that got to newsstands toward the end of the campaign.[33]

But that wasn't all the media! The California recall election also produced an international media feeding frenzy. Hundreds of reporters from all over the world rushed to California in a sort of mini–gold rush. Their presence helped create the furor and drama around Schwarzenegger's every appearance. And it was he they were interested in, not the other candidates or, in most cases, even the recall process itself. Much of the coverage was nothing more than entertainment news gone international and multilingual. Much of it was in a tone of amused condescension: "Those crazy Californians!" But serious journalists came too, and offered serious observations. Here's a sample of comments from the international press[34]:

- "Yesterday's scenes of apotheosis were reminiscent of the happy ending of a Hollywood film. . . . The situation is too complex and difficult for an inexperienced politician and so there is a strong risk that the massive populist revolution he promised will degenerate into disillusionment." (*El Mundo*, Spain)
- "People across the world are in shock: one of America's most prosperous states is to be governed by the Terminator. Hollywood actor Schwarzenegger's victory over his professional governor has once again showed the defective nature of the American electoral system." (*Utusan Malaysia*, Malaysia)
- "fou-fou-fou" (crazy-crazy-crazy). (*Liberation*, France)
- "California has elected a right-wing movie actor as its governor before—Ronald Reagan—but never one so completely inexperienced in the world of politics and government." (*Evening Standard*, London)
- "Someone who's a foreigner in his country, who has an unpronounceable name and can become governor of the biggest American state—that's not nothing." (French Interior Minister Nicolas Sarkozy on French radio)

- "It's the American dream." (Japanese television anchor)
- "This cannot be imagined in China." (posting on a Chinese website)

The Final Cliffhanger

All of this added up to the most unusual media campaign in history. Despite its apparent uniqueness, all the traditional campaign media and techniques were in play. The brevity of the campaign, the open, winner-take-all contest among the replacement candidates, and the star-power of Arnold Schwarzenegger altered the impact of these techniques, but not as much as it changed the role of mainstream news media. While still important, they played a different, less significant role than usual. Schwarzenegger's campaign was particularly successful at going around the mainstream media and directly to the voters, thanks in part to the role played by the nontraditional media of talk radio and the Internet as well as the presence of the entertainment media. But an independent-minded electorate frustrated with politics as usual also played its part.

As Election Day arrived, the mainstream media took the forefront again, ready to report and analyze the results. All three major network anchors—Peter Jennings, Tom Brokaw, and Dan Rather—arrived in California. Pollsters geared up for exit surveys to measure and explain voter behavior. Registrars of voters prepared for higher voter turnout under the watchful eyes of the world media, hoping to avoid a Florida-like fiasco. And the people of California and the world awaited the conclusion of the final cliffhanger.

Part IV

The Outcome

10

Davis Loses the Recall Battle

In the early months of 2003, the state's growing deficit was giving the Gray Davis administration a huge headache. The closer-than-expected, nerve-fraying victory of November 2002 was behind the campaign team now, although not for long. Sadly for the incumbent, the growing demands of the $38-billion budget gap kept the Davis corps from appreciating the shifting winds in the state's volatile political climate. Sure, the administration thought, the deficit was an irritation, but one that could be handled with careful management. Nevertheless, what was characterized as a fiscal concern for state public policy makers would soon become a full-fledged political problem that would ultimately, in a stunning reversal of fortune, cost the governor his job.

Meanwhile, recall proponents were on the move. Even though the reelection had occurred only 3 months earlier, the recall corps believed that the worsening economy would give them their chance to wrest the governorship from Davis. The governor dismissed the effort as the work of poor losers who were attempting to compensate for their inability to win legitimately during the scheduled election: "They couldn't beat [me] fair and square [in November], so now they're trying another trick to remove [me] from office," Davis complained at a February 2003 news conference.[1] Besides, public opinion surveys indicated almost as much voter contempt for the legislature as for the governor. A Field Poll taken in April 2003 showed Davis with a 24 percent approval rating, with the legislature receiving an almost-as-low 31 percent evaluation.[2] Surely the voters would understand the state's difficulties as beyond the control of a single individual, even the governor. Or would they?

They would not. Like the proverbial reference to snowflake distinctions, each election is unique, particularly in a volatile state such as California. And so it was with the Davis collapse, which was the result of a series of converging events—some predictable, others not. In the remaining pages of this chapter, we turn to a few of the leading factors that contributed to the recall of Gray Davis—another colorful chapter in the book of California politics.

A "Normal" Political Campaign in an Abnormal Political Climate

After the recall qualified for the ballot, the Davis team assumed that public opinion would ultimately break along partisan lines. This was cause for hope. After all, Democrats outnumbered Republicans in California by about 10 percent, or 1.3 million voters. Even if Republicans participated in higher percentages, as they tend to do in elections, there would be enough of a cushion to defeat the insurgency as long as voters stayed with their political parties. "We're going to make it clear to all Californians that this thing [the recall] is the handiwork of a little band of right-wing nuts," said Davis adviser Garry South.[3] Simply put, the plan was to cast Davis, state Democrats, and mainstream Californians as moving in the same direction, thus isolating the recall proponents as malcontents. As Davis media strategist David Doak put it, "We found in our research that Democrats come home pretty quick,"[4] implying that partisan alliances would reassemble in time for the election and defeat of the recall.

There was little question about fund-raising for Davis—another good sign. Traditionally, the inability to raise money is an early indication that a candidate is in trouble. But his low standing with the public did not hamper refilling of the Davis campaign coffers. Known as a prodigious fund-raiser, Davis had little trouble collecting enough money to make his case. In all, the candidates would raise more than $83 million over the 75-day campaign, more than $1 million per day. Davis raised his share of funds and then some—more than $20 million went to his campaign alone, which meant that there would be ample opportunity to get out the message. But as is often the case in California, money does not guarantee victory.

Then there was the question of fairness. Even though polls showed voters leaning toward recall, the same surveys found majorities unhappy with the recall as an election instrument. A *Los Angeles Times* statewide survey in late August found likely voters agreeing that the Davis recall set a "dangerous precedent by a margin of 52 percent to 44 percent."[5] Viewed together, the partisan and fairness issues offered a logical path to victory.

Finally, there was the matter of voter uncertainty about what to do. Sure, the voters were angry with Davis, but they could change their minds once the facts came out and cooler heads prevailed. A *Los Angeles Times* poll in late August seemed to confirm voter openness. The survey found that even though a majority wanted Davis removed from office, 46 percent said they could change their minds by the time of the October 7 election.[6] This volatility suggested that there would be plenty of opportunity to turn tentative "yes" votes into firm "no" votes.

Table 10.1

Replacement Candidate Vote

Candidate	Party	Vote	Percent
Arnold Schwarzenegger	Rep.	4,203,596	48.6
Cruz M. Bustamante	Dem.	2,723,768	31.5
Tom McClintock	Rep.	1,160,182	13.5
Peter Camejo	Grn.	242,169	2.8
Arianna Huffington	Ind.	47,486	0.6
Others		276,458	3.0
Total			100.0

For all of these reasons, the Davis people remained confident that they would prevail by the time of the election. Politics has a rhythm, and whatever the ruckus of the moment, that rhythm would restore political order in due course. Yet, as the summer turned to fall, it became increasingly clear that voter anger was not about to diminish. The benefits of incumbency, party loyalties, traditional fund-raising, and public concerns about the process notwithstanding, Davis would have to go.

A Mandate for Change

The voters spoke with amazing clarity on Election Day. On the question of whether Governor Davis should be recalled, the electorate said "yes" by a margin of 55.4 to 44.6 percent. All of the Davis endorsements, appearances by national Democratic heavyweights, and decades of political experience were pushed aside in favor of new leadership. But by whom?

On the question of who would replace Davis, the first choice of the voters was Arnold Schwarzenegger. No, Schwarzenegger did not get a majority, but he came close with 48.6 percent (Table 10.1). That percentage was remarkable, given so many other replacement candidates on the ballot. There were other noteworthy comparisons. Schwarzenegger's percentage of the replacement vote was greater than the percentage of "No on Recall" votes for Davis (45 percent), which was the question on the first part of the ballot. Schwarzenegger also received a higher percentage than that received by Davis in his defeat of Republican Bill Simon in 2002 (47 to 42 percent).

If election "mandates" are defined by majority votes, Schwarzenegger missed the mark—but barely. By any other measure, however, he clearly emerged as the people's choice. Moreover, combined, Republicans Arnold Schwarzenegger and Tom McClintock received over 60 percent of the vote; that total most certainly spoke volumes about mandates and change.

Key Factors

Just as a series of unique conditions combined to place California in an economic straightjacket prior to the recall effort, so it was that several factors blended to forge an incredible election outcome on October 7. Together, they made the Davis demise one-sided and complete.

Davis and the Issues

Although the budget deficit was foremost in almost everyone's mind, other issues surfaced during the campaign to bring on further problems for Davis. One had to do with the annual motor vehicle fee paid by car owners, based on the value of their cars. When the state budget was flush during the late '90s, the legislature reduced the fees by two-thirds, with the proviso that should the economy go into a tailspin, the old levels would return. As the deficit soared in early 2003, the governor proposed various combinations of tax increases along with spending cuts; while the legislature accepted much of the latter, Republican members blocked the former.

With little movement in the legislature, Davis felt he had no choice but to restore the motor vehicle tax at the old levels—there was little else he could do without the cooperation of the legislature. The voters fumed, and every major recall candidate vowed to repeal the increase. After the election, the governor claimed that Republican Senate Minority Leader Jim Brulte reversed an earlier promise to raise taxes as part of a deal for closing the budget gap because of the growing success with the recall petition effort.[7]

Immigration, always a testy issue in California, also found its way into the campaign when Davis promised to sign a bill allowing illegal immigrants to apply for drivers' licenses. The new law was considerably weaker than an earlier version that required background checks and which Davis had vetoed. With his about-face, Davis incurred the wrath of critics, who accused the governor of pandering to Latino voters,[8] underscoring the belief that to Davis, expediency was more important than values.

Davis and Endorsements

Several key supporters from past elections were conspicuously quiet or absent from the Davis camp in the 2003 recall election campaign. Unions such as the California State Employees Association divided their campaign contributions and phone banking between Davis and replacement candidate Cruz Bustamante. The state prison guards' union, a huge supporter of Davis in 2002, remained eerily silent. Education groups seemed to operate at only half speed, missing the vigor characteristically so much a part of their efforts. Only core party leaders and elected Democrats seemed to speak out vigorously on behalf of Davis. Yet those Democrats who spoke most passionately for Davis were mostly out-of-state leaders such as presidential hopefuls, former President Bill Clinton, and civil rights activist Jesse Jackson. Only state Democratic Party Chair Art Torres was consistently seen at Davis's side; other state leaders played muted roles.

Then there was the Hollywood community. Traditionally, people in the entertainment field are disproportionately liberal and Democratic. In star-struck California, these big names are magnets for money, votes, and media attention. Maybe so, but this time around celebrities Rob Lowe, Kelsey Grammer, Dennis Miller, and Tom Arnold tossed their support to Arnold Schwarzenegger, a telegenic counterpart to the dry Davis. Other long-time Democrats such as Barbra Streisand, Dustin Hoffman, Larry David (*Seinfeld* creator), and Aaron Sorkin (*West Wing* creator) went for Arianna Huffington as a replacement candidate, although they described themselves as members of the "No on Recall" camp. Unlike most elections, this time there would be no Democratic lock on Hollywood—not with a Republican action hero running hard.

Davis versus Davis

With more than 30 years in the public eye, Davis was hardly an unknown to the voters. Yet concern about Davis "the person" became central to the recall effort. Political allies often referred to the governor as intelligent, disciplined, and extraordinarily capable of governing the world's sixth largest economy. Wife Sharon Davis repeatedly attempted to portray her husband as a warm, caring individual, with the couple having "the same likes and dislikes as other Californians."[9] This small circle aside, the numbers of Davis defenders were few.

Rather than reach out to people, Davis stayed on messages framed in numbers and values. Outside of fund-raising, it was hard for Davis to mix with people. As one account stated, "Davis was not a schmoozer. He didn't glad-hand with people who couldn't help him."[10] His stiffness carried all the way to his attire. Said a prominent L.A. clothier, "Gray Davis has no style. I'd like to see him in a closer-fitting suit, some shirts with more color and ties that make some sort of statement rather than melting into the woodwork. . . . And he looks like a 15½-inch neck wearing a 16½-inch shirt."[11] A superficial analysis? Perhaps, but in an image-conscious state like California, Davis's attire and body language seemed to speak volumes about his personality and character. The voters agreed. According to a Field Poll conducted in August

2003, a stunning 68 percent of those interviewed stated that they disliked Gray Davis as a person.[12]

Davis realized as much, but too late to do anything about it. In a post-election analysis of the recall experience, the governor said that not mixing with voters in town hall meetings and similar venues was a major mistake not only of his campaign but of his governorship. "It's worked for [President George W.] Bush," he said, "and it would have worked for me."[13]

So why should personality matter anyway? When the economy is healthy and people are content, they care little about who governs. But when things go sour, people need a reason to believe that circumstances will improve and a leader to take them to better times; should those elements fail to surface, they will look for someone to blame. That's where "personality politics" comes in. By emoting in terms that comfort the public, Davis could have bonded *with* the voters instead of appearing removed *from* them. Yet it was not to be. Entrusted with leading the state to better times, Davis simply did not enjoy the public's trust or inspire confidence. As *San Francisco Chronicle* reporter Robert Salladay concluded, "A good politician who speaks to the masses can get away with amazing things. But Davis got no forgiveness for his questionable fund raising, no forgiveness for a budget deficit amid a national recession, no forgiveness for an energy crisis manipulated from Texas."[14]

Because he seemed so aloof from the people, voters didn't give Davis the benefit of *any* doubt. In the words of a leading Democratic strategist, "You have the same trends [in California] that are facing probably 48 governors in the country that blew a hole in the budget and caused other economic problems. Because Gray had no personal capital to work with, he had no cushion to fall back on."[15] For these reasons, Davis's personality became a critical factor in the recall.

The Schwarzenegger Surprise

By the end of September, the recall had settled basically into a two-person race. Public opinion surveys consistently showed that if the voters sent Democratic Governor Gray Davis packing, they would choose Republican Arnold Schwarzenegger as his replacement. Cruz Bustamante, the only well-known Democratic replacement candidate, had faltered badly, although he remained a thorn in Davis's side to the extent that many of his supporters might vote "yes" on the recall in an effort to position the lieutenant governor to win.

Initially, the partisan division seemed to favor the governor, but Schwarzenegger proved to be an unconventional Republican target. Conservative on taxes and spending, he displayed moderate to liberal tones on abortion, immigration, gun control, and public education—approaches that appealed to most Californians. Moreover, Davis could not get the disciplined Schwarzenegger to deviate from his campaign of reaching out to the people directly and through national interviews on shows such as the *Oprah Winfrey Show*. With the race now simplified, Davis called for a one-on-one debate, but Schwarzenegger would have none of it.[16] With time running out, Davis's hopes seemed to be fading. Then came the issue.

Less than a week remained in the recall campaign when the *Los Angeles Times* reported the Schwarzenegger groping allegations. Schwarzenegger immediately issued an apology for his past behavior; at the same time, however, he also castigated the *Times* for last minute "trash politics." And with that, the Schwarzenegger campaign moved on.

With a reputation for "slash and burn" campaigning, Davis knew he had to play the Schwarzenegger issue carefully. Even ally Democratic Attorney General Bill Lockyer had publicly warned Davis not to do the "trashy campaign" that he had carried out so effectively against former Los Angeles Mayor Richard Riordan during the Republican primary in 2002.[17] At first, Davis stayed clear of the *Times* allegation, saying only that on October 7 "the voters will determine how significant that story is."[18] But time was quickly running out for Davis, and 24 hours later he reversed course. "If the allegations are true," Davis said at a campaign rally, "I believe that he [Schwarzenegger] is morally unfit to hold office."[19] Davis also suggested that Schwarzenegger may have broken the law. These remarks marked the last stage of the ill-fated Davis campaign.

The groping stories didn't resonate with the public beyond a "so what?"; therefore, Davis was left campaigning on what was quickly reduced to a nonissue for most voters. In fact, an NBC exit poll taken on October 7 found that those people who made up their minds during the last week of the campaign voted for the recall by a 57 percent to 43 percent margin, slightly higher than the overall pro-recall vote. Even among female voters, the allegations held little sway as a factor in their decision.[20]

Voting Groups

The Davis team hoped for a high turnout in the special recall election. This was a highly unlikely development, inasmuch as special elections—elections without the full panoply of contests to bring out different voters with interests in different races—usually produce abysmally low turnouts. Not so in 2003, where 61 percent of the registered voters cast ballots, significantly higher than the 50 percent turnout in 2002. In normal times, such a bump would have benefited Democrats, but nothing was "normal" about the 2003 recall election.

Virtually every element of the Davis coalition from 1998 and 2002 provided less support for Gray Davis in the 2003 recall election. Exit polls taken for NBC showed that a full quarter of all Democrats voted to recall Democrat Davis, twice the usual defection rate. This was perhaps the most damning of all election night data. As if to add insult to injury, about 20 percent of Schwarzenegger's vote came from Democrats.

The strategy of both Davis and Bustamante had been to shore up their base, counting on the loyalty of Democratic voters, who substantially outnumber Republicans in California. This strategy led both Davis and Bustamante to move to the left on policy issues and legislation. But the strategy failed. In moving to the left, they ceded the center to Schwarzenegger, who eagerly claimed it. And much of their Democratic base abandoned them. Fully 60 percent of the voters chose Republican candidates Schwarzenegger or McClintock.

With respect to ethnicity, 47 percent of Latinos voted for the recall, leaving only 53 percent voting against it and for Davis. This support was markedly lower than the 78 percent Latino vote that went to Davis in 1998 and the 65 percent share he received in 2002. Moreover, Latinos accounted for a record 17 percent of the vote in the recall election—a greater proportion of the electorate than ever before. The Davis team had hoped for a high Latino turnout, but these voters, like so many others, marched in a new, independent direction. African Americans, on the other hand, were the most loyal group, with 79 percent opposing the recall—a substantial number but still lower than the percentage Democratic candidates can usually count on. Sixty-five percent voted for Bustamante—higher than his 55 percent from Latino voters. Asian voters, however, reflected the larger electorate, with 53 percent supporting recall and 45 percent choosing Schwarzenegger.[21]

The gender gap between Democrats and Republicans narrowed and almost disappeared in this election. Democrats usually get close to 60 percent of the women's vote, thanks to their positions on abortion, gun control, and education. But in this race, 51 percent of women voted for the recall and 59 percent supported Schwarzenegger or McClintock, all but eliminating the famous "gender gap." Gay, lesbian, and bisexual voters were more supportive of Davis, however, with 58 percent voting "no" on recall.

Sizable numbers of union voters also deserted Davis. Union members represent 18 percent of the state's work force, a much larger fraction than the 13 percent figure nationwide. Historically, they have been loyal to Democrats in general and Davis in particular. In 2002, union members went for Davis over Republican Bill Simon by a margin of 3 to 2. In 2003, however, union members split their votes almost down the middle, defying the wishes—and endorsements—of their leaders. Another crack in the Democratic coalition.

Figure 10.1 **2003 California Statewide Special Election: Shall Gray Davis Be Recalled? Counties Voting "Yes" or "No" by Percentage** (the number in each county indicates the percentage by which the majority of voters voted "yes" or "no" in that county)

Source: California Secretary of State.

Davis had problems with other constituencies, too. In terms of geographic regions, as Figure 10.1 indicates, only the San Francisco Bay area vigorously supported Davis; all other areas voted for the recall. While Los Angeles County voted against the recall by a slim 51-to-49 percent margin, the Davis camp knew they needed a much more decisive vote in order to offset the conservative areas of Southern California and interior regions. Interestingly, while the economy and jobs topped the list of voter concerns, the Bay Area, with the highest level of unemployment in urban California, opposed the recall, while Southern California, much of which had not been so hard hit by recession, supported recall.

As for education and income, recall proponents prevailed in almost every category. Almost every age group, too, went against Davis. Somewhat surprisingly, however, Schwarzenegger didn't do better with young voters and new voters than with other groups. Many observers expected that his star power would draw lots of new voters into the political process, but exit polls indicate that it did not. It was also thought that he would appeal to younger voters, but those aged 18 to 29 were actually a little less inclined to support recall and Schwarzenegger than other groups.

Despite some variation and a few Democratic strongholds like African-American voters and voters the San Francisco Bay Area, almost every demographic group went against recall and for Schwarzenegger. Just 11 months after they had reelected him, California voters sent Governor Gray Davis a clear message that he was no longer welcome. That message was inescapable. But was there a more general message to Democrats? Or was it a message about politics as usual?

A Smooth Operation After All

Prior to the shortened election cycle, state and local voting officials worried that the short preparation period, antiquated voting machines, and reduced number of voting stations would contribute to a gridlock on Election Day resembling rush hour on California's infamous freeways. Davis, the American Civil Liberties Union, the National Association for the Advancement of Colored People, and other groups had seized on these concerns and filed suit to delay the recall election until the scheduled March 2, 2004, primary election.[22] No doubt some of the concerns were political—the Davis team believed that a March election would bring out a larger turnout of Davis supporters.[23] Part of the concern, however, was truly based on the logistics of the old punch card system. One study by a professor at U.C. Berkeley had found the likelihood that as many as 1.6 percent of punch cards (the system used in the most populated counties) would be invalidated because of the "hanging chads" that had become so infamous in Florida in the 2000 presidential election.[24] As we already know, state and federal courts turned away these challenges. But were they right to do so?

It turns out that the election went smoothly after all. One exit poll on Election Day found that only 2 percent of the voters experienced serious problems with voting equipment or the length of the ballot. Moreover, those who used punch cards were no more likely to encounter difficulties than those who used the modern electronic voting systems.[25] Antiquated or not, the system held up just fine on Election Day. No doubt some of the potential waiting problems at voting sites were mitigated by the fact that a record 30 percent of the electorate cast their votes via absentee ballot.

Anatomy of a Defeat

Often, in reflecting on an election outcome, observers seize on that pivotal moment when the election changes course. Once the recall drive got under way, there was no such single moment for Gray Davis. What started out bad became worse over time. In the voters' eyes, Davis had dug himself into a hole so big that he would never get out.

Operating with little public trust, Davis had precious little time and less opportunity to reverse course and undo all the bad will that had built up. Whenever he signed a bill, growing numbers of detractors publicly doubted his motives. Still the question remained: How could the voters reverse their own course from just 11 months earlier over such a short period of time? Here, too, there was no immediate answer or the appearance of a single revelation. Rather, the steam of the state's collective pressure pot seemed to build so much that only a complete venting of the voters' anxiety would restore any sense of calm. Whatever their reasons, the voters spoke loudly and in a clear voice on October 7.

11

Recall and Political Stability in California

Can They Coexist?

On October 7, 2003, the California electorate decided to remove Gray Davis from his job as governor. Upon doing so, they chose actor and former body-builder Arnold Schwarzenegger as their new governor. With those two bal-lots, the voters reversed their decision of 11 months earlier and put into office someone with no significant previous political experience.

The fact that replacement candidate Schwarzenegger received considerably more votes than the number of votes to keep Davis in office eased some of the concerns about the legitimacy of the election. Given the circus-like atmosphere of 135 candidates and a whirlwind campaign, in many respects it was amazing that anything definitive emerged. Yet, the people had spoken on the replace-ment issue with amazing clarity. There are other questions about the recall event, however, and on these the answers may not be so clear.

What is the significance of this political rupture and the conditions under which it occurred? What does the outcome say about the future of California politics? Does the recall portend a new direction for the rest of the nation, or is it yet another example of "wacko" California at its best?

Perhaps it is too early to know the answers to these questions. Neverthe-less, in the remaining pages, we focus on lessons learned, lingering struc-tural problems in California politics and government, and the question that started this discussion—namely, whether the events precipitating the recall combined to be a unique "perfect storm" or whether they underscored California's status as the nation's trendsetter.

Putting the Pieces Back Together

Just days after one of the most acrimonious election events in recent memory, the transition from the administration of Gray Davis to that of Arnold

Schwarzenegger began in earnest. Swiftly, pre-election combat became post-election cooperation. Schwarzenegger journeyed to Sacramento for first-hand briefings on the state's issues and conferences with state legislative leaders. But perhaps the most poignant moment of all came on October 23, when the outgoing and incoming governors met to discuss the state of the state.

In their 90-minute meeting, Davis and Schwarzenegger covered both challenges and opportunities confronting the state. At the end of their conference, Davis handed Schwarzenegger a massive white notebook entitled "Transition 2003." They next briefly addressed the press. "I'm going to do my very best to help Governor-Elect Schwarzenegger because I love this state," said a wistful Davis. Countered successor Schwarzenegger, "I have to say that my work in the future will be helped by Governor Davis because . . . he's going to give me some of the inside information and he's going to help me with the transition."[1] With that exchange, the two former adversaries set the tone for the changeover. Whatever the pain from lingering wounds, whatever the exhilaration of victory, the governance of California had to move forward.

Lessons Learned

Davis did his best to put the disastrous defeat into perspective. After all, he was only the second governor ever to be recalled from office. Nevertheless, taking a page from Schwarzenegger's successful national television exposure on numerous programs, Davis appeared on *Late Night* with David Letterman. There the outgoing governor offered his "Top 10" pieces of advice for his successor.

10. "When you realize you don't know what you're doing—give me a call."
9. "Baby oil will stain the [governor] mansion's Italian silk sofa."
8. "Listen to your constituents—except Michael Jackson."
7. "Sorry. Joke Number 7 was recalled."
6. "To improve your personal ratings, go on Leno. When you get kicked out, go on Letterman."
5. "Study the master, George W. Bush. Ah, I'm just kidding."
4. "You could solve the [$38 billion] deficit problem by donating your salary from *Terminator 3*."
3. "If things are bad, just yell 'Save us, Superman.'"
2. "While giving a speech, never say, 'Santa Cruz, Santa Barbara . . . same thing.'"
1. "It's pronounced California" (in reference to Schwarzenegger's Austrian accent, which led him to pronounce the state's name as "Kallee-*foh*'nia.")[2]

Such humor marked a rare moment in what had been a tense and no doubt painful year for Gray Davis. Nevertheless, in the aftermath of one of the most bizarre chapters in California politics, there are a few observations worth noting. Some confirm past behaviors; others suggest that the state may be moving in a new direction. Regardless, we need to bear in mind the stamp of the 2003 recall.

Drowning in an Economic Funk

If you ever doubt that people are affected by budget imbalances, just look at the impact of the state's deficit on the popularity of Gray Davis. As the deficit waxed, his prestige waned. In July 2003, a Field Poll found 75 percent of the respondents believing that the state was on the "wrong track." In the same poll, 61 percent placed the blame on Davis for not resolving the state's budget deficit, whereas 47 percent reproached the legislature.[3] Over the next 3 months, matters went from bad to worse. According to exit poll results on Election Day, 83 percent of the respondents believed that the state's economy was in bad shape, compared with 56 percent of a national sample who answered the same question.[4] Moreover, by an overwhelming margin of nearly 4 to 1, the voters put the blame for the state's problems squarely on the shoulders of Gray Davis.[5]

The Power of Celebrity

Rather than get caught up in unscripted debates and face scrutiny by other candidates and the media, Schwarzenegger used his celebrity to bypass potentially troublesome confrontations. He "sold" his candidacy directly to the voters by treating them as fans. And it worked. In the words of one account, "Jay Leno and Oprah Winfrey replaced sober questioning from editorial boards. Howard Stern had more resonance [with the voters] than the learned opinions of California's newspaper columnists."[6] Who wanted to read boring accounts of candidate exchanges in the newspapers when they could watch Schwarzenegger mixing it up with the hosts of *Entertainment Tonight* or *Access Hollywood*? Simply put, Schwarzenegger waged his battle for the governorship on his turf and on his terms. And he reached the voters on their medium of choice—television. As Sean Walsh, co-director of communications, put it, "we [the Schwarzenegger campaign] ran away from the established media. We make no apologies for doing lots of radio or TV. It gave us 5, 7, 8 minutes of unfiltered opportunities to get out our message every day. We did it because we could."[7]

Personality versus No Personality

The comparison of Gray Davis with Arnold Schwarzenegger became a painful mismatch over the course of the campaign. Davis exhibited the image of a cautious, wooden figure who was most comfortable spouting policy-wonk figures. Because of his obsession with the micromanagement of staff, he insisted on keeping his hands on the controls of government, thereby excluding others. Wrote *Los Angeles Times* political columnist George Skelton, Davis "wouldn't schmooze. Worse, he was insensitive and rude."[8] Schwarzenegger couldn't have been more different. With a penchant for slick lines from his movies rather than mundane statistics about the state's economy, Schwarzenegger said the people didn't care about data and details. When asked in a rare press conference to provide specifics in addressing the state's fiscal crisis, Schwarzenegger exclaimed, "What the people want to hear is: Are you willing to make changes? Are you tough enough to go in there and provide leadership? That's what this [election] is about."[9] Maybe it was, at least in part. Regardless, the comparison of Davis with Schwarzenegger made the contrast all the easier to see.

The Permanence of Structural Issues

Irrespective of who emerged as the victor in the recall election, the morning after would not be pleasant. California is a state in deep financial trouble, and changing governors in itself won't do much to cure the state's major ills. Other structural impediments to smooth policy making will not go away without major reorganization. Given such a framework, unmet expectations may cause even more problems down the road. But for now, let us remain in the present.

Independent Political Offices

Arnold Schwarzenegger's victory gave him something close to a mandate, but all the other statewide officeholders with whom he was to share power were Democrats, and most were eager to run for governor. Independently elected, each had no compunction about relying on the constitutionally defined power of his office, even with Democrat Gray Davis in the governor's office. Davis and Lieutenant Governor Cruz Bustamante had problems, but at least they were members of the same political party. Schwarzenegger, however, immediately faced the challenge of running a state with his "number two" as a man who ran against him. And when the governor leaves the state, the lieutenant governor is in charge. . . . "Arnold, you're very famous for making movies all over the world," Bustamante said after the election. "And I want you to feel free to continue doing that. Go where you like. Feel free to stay as long as you like. I'll be here, keeping an eye on things."[10]

Unstable Revenue Bases

There is a reason that California is known for its "boom and bust" economy: the state's tax structure moves with greater swings than the dial on a seismograph calibrating the intensity of an earthquake. Such a shift occurred between 2000 and 2002. In that brief 2-year period, the state's receipts from income taxes associated with capital gains and stock options dropped by 70 percent, from $17 billion to less than $5 billion. The revenue loss was compounded for two reasons: California relies on the income tax for about half of its general budget revenues, and the progressive nature of the tax (the more you make, the higher the percentage you pay) has resulted in 10 percent of the filers paying about 75 percent of all income taxes. So when the dot-com boom went bust, so did much of California's income from a select few filers. Ironically, Davis wanted to solve the shortfall by actually adding two higher income brackets.[11] While this change might have closed the gap for the moment, it would have contributed to even greater economic swings in the future.

Other sources pose problems for different reasons. Because of California's Proposition 13, property taxes make no allowances for taxing businesses at different rates than residences, nor do they allow for the increasing value of real estate over time. Sales taxes are problematic because of numerous exemptions, leaving California with a relatively narrow tax base compared with other states. Thanks to vigorous interest group activity, corporate taxes are low and riddled with enough loopholes to make Swiss cheese look whole. Combined, the weaknesses of these taxes left California coffers unusually dependent on the income tax—a tax that suddenly proved itself unreliable as a revenue source.

Inherent Difficulties of a High Legislative Voting Threshold

California is a tough state to govern, in part because of its rules. The requirement of an absolute-two-thirds vote on the budget sets up one of two undesirable scenarios: either one party dominates the legislature by a lopsided margin, thereby rendering the other party impotent, or the balance between the two parties leads to gridlock. Because of the absolute two-thirds rule, the legislature has been unable to pass the budget on time during 17 of the past 25 years. Missing the mark is no big deal when the economy is humming, but when major problems exist, as during the period between 2001 and 2003, the public is much less patient. Sure, the legislature has responsibility for passing the budget—at least in theory. But that's "in theory." Remember, no one is more visible in a state than the governor. Thus, right or wrong, Gray Davis received much of the blame for the legislature's budgetary problems.[12]

Reform could be on the way, however. An initiative to reduce the requirement of an absolute two-thirds majority to approve the budget or raise taxes qualified for the March 2004 ballot. In another manifestation of California's diversity and complexity, the measure was sponsored by a coalition of public employee unions. Public opinion polls in 2003 reported voters unwilling to lower the two-thirds requirement, but if it is not changed in 2004, it surely will continue as an issue—and problem—in its own right in California politics.

Term Limits

Let's get one thing straight: state legislative term limits didn't drive Gray Davis from office. After all, the concept of term limits terminates the ability of a legislator to hold office beyond a constitutionally fixed period of time. So what is the relation of term limits to the recall? It's simply that term limits have stripped the legislature of the experience so necessary to organize a complicated state budget; without expertise, cultivated personal relationships, and the know-how of compromise, legislators tend to take and stay with rigid ideological positions.

Fast-forward to the recent budget stalemates and you see a legislature that wasn't about to move on the budget in any cooperative way, leaving Gray Davis incurring the wrath of the voters as a consequence. As long as we have strict limits on legislative service, legislators will be both beholden to the governor and unlikely to coalesce in any effective way on potentially divisive issues. So if they're beholden to the governor, why didn't the legislature cave in to Davis's demands? In this case, the poor relationship between legislators and the governor exacerbated the isolation of each branch, leaving—you guessed it—political intransigence.

Political insiders and academics talk about the problems caused by term limits, and reform proposals are occasionally discussed, perhaps extending service in the assembly to 8 years and in the senate to 12 years, for example. But the term limits are sacred to many voters, and until the approval ratings of the legislature improve, reform is unlikely.

Flawed Redistricting

Under California's constitution, the legislature has the responsibility for redistricting after each census. The idea is to make sure that each legislative district contains approximately the same number of people. As long as they go "by the numbers," the legislature can use redistricting as a vehicle for protecting its members with "safe" seats, rather than creating competitive districts where moderates would most likely prevail over extremists. That happened in 2001. As reported in the *California Journal*, Democrats and Republicans agreed to a redistricting plan that protected the status quo. "Democrats, who may have been able to add a few more seats with more creative mapmaking, agreed to the deal in order to head off a potentially lethal Republican challenge in the courts. Republicans went along because the new scheme kept their numbers from eroding further and because Democrats controlled the entire redistricting process—including the governor's office."[13] With so many legislators coming from "safe districts," there is little reason for anyone to work with anyone else on thorny public policy issues.

The issue has not gone away, even in the post-recall period. In November 2003, recall leader Ted Costa began circulating an initiative petition that would have post-census redistricting conducted by a group of retired judges without partisan backgrounds.[14] If Costa doesn't do it, Schwarzenegger might.

Weak Political Parties

The weak organizational structure of California's political parties further blurs the lines of governance and invites unnecessary chaos. Unlike those of many other states, California's political parties do not have the ability to control nominations.[15] The fragmented party system revealed itself in 2003 on both sides of the recall issue. That Democratic Lieutenant Governor Cruz Bustamante ran as a replacement candidate against fellow Democrat Gray Davis was a blow to the hope for Democratic unity. His "No on Recall, Yes on Bustamante" campaign was incredibly transparent. The fact is that Bustamante had no hope of winning the replacement election unless people voted "yes" on the recall, and that's why he became a candidate. As a result, some Democrats voted against Davis to choose Bustamante. Democratic-leaning funding sources were conflicted about the twin candidacies and just didn't know what to do. So fractured were the Democrats that after the recall election, Democratic Attorney General Bill Lockyer announced that while he voted against the recall, he also voted for Republican Arnold Schwarzenegger.[16]

Matters were equally confusing for the Republicans. The age-old schism between party conservatives and moderates resurfaced in the replacement race, with moderate Arnold Schwarzenegger and conservative Tom McClintock each appealing to core constituencies, along with other Republicans in the race. One by one, leading Republicans endorsed Schwarzenegger and asked McClintock to drop out in the name of "party unity," but they had no way of forcing him from the race. As with so many other times in the past, the division could have allowed either Davis to retain his position or Bustamante to win the replacement race by virtue of being the only "top-tier" Democrat. Only massive contempt for Davis and—by association—Bustamante permitted Schwarzenegger to prevail.

So why should we care if the political parties are weak and ineffective? The fact is that whatever their imperfections, in democracies political parties remain the most important mechanisms for sorting and representing political values.[17] In the process of fulfilling these tasks, party labels help us identify with those seeking office. The meanings of those labels are much more difficult to interpret when many candidates from the same party are running at the same time. The two-part recall and replacement election operated under such circumstances.

But the recall revealed something else about California voters and political parties. In the past, with governors like Earl Warren (1943–1953), Ronald Reagan (1967–1975), and Jerry Brown (1975–1983), California voters were thought to focus more on the personalities of the candidates than on their political parties. Instead of consistently voting for candidates of one party or the other, they commonly switched back and forth, not only between elections but even on the same ballot. Then, in the 1990s, the voters became far more consistent and political scientists suggested that the era of personality politics in California had ended. The recall election, however, suggests otherwise. In a state where Democrats outnumber Republicans by 1.3 million, Republicans Schwarzenegger and McClintock won 60 percent of the vote. Was the state shifting to the right? Perhaps, but even as these two were winning 60 percent, the "no" vote on Proposition 54, the racial privacy initiative, was 64 percent! Hardly the product of a right-wing electorate.

Two other explanations seem more credible. Clearly people were drawn to Arnold Schwarzenegger by his personality—and repelled from Gray Davis by his. So perhaps personality politics was back at work. With Arnold Schwarzenegger in the race, it's hard to argue otherwise. Alternatively, however, the vote could be explained by Californians choosing the moderate option, as they often have in the past. As Gray Davis and Cruz Bustamante painted themselves into a liberal corner and Tom McClintock martyred himself on the alter of conservative purity, Schwarzenegger was left in the middle. And the middle proved to be big. California's shift to the Democratic Party in the 1990s may have resulted more from the archconservative candidates offered by the Republican Party than from the policies of their opponents. When Republicans finally offered a candidate who could be seen as moderate, they won.

But California's primary election system makes it hard for moderates to win their party's nomination, because primaries are dominated by party loyalists, and they tend to be more conservative or more liberal than mainstream voters, as Dick Riordan learned in the 2002 Republican gubernatorial primary. "I don't believe for a second that Arnold Schwarzenegger could have won a Republican primary last year," said a former chairman of the California Republican Party.[18] Without a primary, a moderate like Schwarzenegger could get past the ideological purists and present himself to the full electorate—which, it turns out, is more centrist than many Democrats have been assuming.

The combination of primary elections, highly partisan districts, and weak political party organizations constitutes another problem for Governor Schwarzenegger: leading his own party in the state legislature. Because of these electoral structures, most Republican legislators are more conservative than Arnold Schwarzenegger. They're independently elected from secure districts, which may make them reluctant to follow the more moderate governor. Schwarzenegger must decide whether to attempt to remake his own party by intervening in primary elections—which many moderate Republicans hope he will do—or accept the status quo and look to Democrats for compromise rather than legislators of his own party.

All of which, combined with the lesson of the recall, could result in yet another reform. In 1996, California tried a blanket (or "open") primary system that allowed voters of any party to vote for candidates of any party in the primary election, crossing over from their own party to another if they wished. Advocates of the open primary argued that it would produce more moderate candidates, more competitive elections, and greater willingness to compromise in the legislature. The courts, however, overruled the initiative on grounds that it interfered with the freedom of political parties to choose their own candidates. After the recall, advocates of the open primary were considering ways to revive it, again through the initiative process.

The Future of California Politics

"This election was the people's veto—for politics as usual," Governor-Elect Schwarzenegger declared in his swearing-in speech. Kevin Starr, California's most noted historian and the state librarian, seemed to concur, declaring that "The crisis requires nothing less than the refounding of our state." Reforms have been proposed for each of the structural problems discussed above. The question is whether Californians will have the will to carry any of them out—and whether Governor Schwarzenegger will choose to lead the state through any or all of the reforms.

Clearly he has the potential to do so. If governors have the advantage of the bully pulpit, no gubernatorial pulpit has been bullier than Schwarzenegger's. In the campaign and upon his election, he had unprecedented access to the public through his ability to command the attention of the media. In the weeks immediately following the election, he used that access to push his policies and follow through on his promises. In years past, the media have been withdrawing from Sacramento and coverage of state politics. No television station in the state except those in Sacramento had bureaus in the state capitol, and even newspapers had been reducing staffing in the capitol. But when Schwarzenegger was elected, the media returned en masse. The capitol scrambled to accommodate them and even the national networks announced plans to set up Sacramento bureaus.

Governor Schwarzenegger made it clear that if he couldn't get the policies he wanted from the legislature, he would take the issues to the voters through initiatives. From most governors this would have been a largely idle and distinctly limited threat. From Schwarzenegger, with his media access, it was credible. The combination of such a leader, so much media attention, and the mechanisms of direct democracy could result in a very different kind of politics and government.

But Governor Schwarzenegger faces another sort of problem. Part of the challenge of governing California lies in its complex diversity, which makes it hard to bring Californians to agreement. A particular challenge will be reaching some sort of consensus on what level of services Californians want and what price, in taxes, they are willing to pay for it. For all of California's problems are not structural. Some of them are in the very nature of the state itself. As Walt Kelly's cartoon character Pogo used to say, "We have met the enemy and he is us."

California Recall: "Perfect Storm" or Trendsetter?

Was the recall of Governor Gray Davis by the voters of California the result of "perfect storm," where several normally unrelated conditions converged to move the state in an unprecedented manner? Or was this huge political occurrence the beginning of a new trend, where restless voters will act on their own schedule rather than the traditional election schedule? The answers to these questions could well serve as a forecast for American politics in the coming years.

The "Perfect Storm" Scenario

Clearly, Gray Davis brought on neither the power crisis nor the serious recession that combined as one-two punches between 2000 and 2003, rocking the state in their wake. The first event was caused largely by poor planning before Davis's watch, while the latter resulted more from a national economic implosion than anything that could have arisen in California alone.

What about the budget deficit? Here the evidence is not so cut and dried, for although Davis, as governor, did not bring on the economic malaise responsible for shrinking revenues, he and his advisers saw the deficit coming 2 years before it hit with full force. State government expenses could have been cut before the gap grew so big, but they weren't. Yes, Davis had help from a timid legislature, which—like the deer staring into the headlights—seemed too paralyzed to move. Nevertheless, as the state's highest-ranking elected official, it was Davis's duty to sound the fiscal alarm as soon as possible, and he did not. Whether it was because the governor expected the dip to reverse much earlier or because he did not want to bring bad news to the electorate before the 2002 election, Davis missed the boat.

Finally, there is the double-barreled question of the governor's character and personality. When things are going well, the public doesn't take much time to consider these concepts because they are, well, squishy and seemingly unrelated to daily routines. But when trouble occurs, these elements take on new significance. As a hard campaigner, Davis was sometimes criticized for what opponents referred to as "pay-to-play" politics—meaning that he wouldn't listen to people and/or groups unless they first contributed to his campaign. Thus, right or wrong, he had the reputation in some quarters as an individual who cared more about the outcome than any core values. The other aspect, his personality, seemed cool, calm, and deliberative when the state fared well but cold, calculating, and divisive when the economy soured. Which was the real Davis? The answer depended on whom you asked and the conditions at the time. Then, again, maybe they were one and the same. If that seems fickle, perhaps it is, but it's also the reality of politics. Perhaps *Los Angeles Times* columnist George Shelton summarized Davis's problems best when he wrote that Davis "lacked three assets crucial for any California governor. They are long-range vision, core convictions and people skills. A governor can survive without one, maybe two, but not all three."[19] Still, Davis committed no crime.

Add to all these factors the availability of the recall as one of the tools of direct democracy in California and, with the complex diversity of the state, you have the breeding ground for a little band of passionate activists to launch a campaign that at first looked like tilting at windmills. Finally, add the infusion of nearly $2 million for gathering recall signatures, and you have a set of unusual and impressive conditions that seem unlikely ever to be replicated, suggesting that the recall was indeed a unique event.

The Trendsetter Scenario

All this takes us to the second possibility, namely that the recall is the beginning of a new trend no different from others—bilingual education, medical marijuana, term limits, and property tax reform, to name a few—that have been spawned in California and spread across the nation. There may be something to such an idea. In 2003 alone, voters in Arizona, Nevada, Montana, and Wisconsin circulated petitions to remove their governors. Each of these states had qualification thresholds higher than California's, and efforts in all failed. Still, anecdotal information suggests that voter impatience with elected officials may be on the rise, with the ability to organize a recall election serving as a way of venting those frustrations.

Some people agree. M. Dane Waters, founder and president of the Initiative and Referendum Institute, views the California experience as a potential turning point for new voter activism: "Before the recall drive in California, I had one call every six months. In the past two months, I've had calls from all over the country asking whether they can recall elected officials in city halls and state legislatures."[20] Moreover, it's been more than anecdotal in Wisconsin where, in September 2003, a state senator was recalled for only the second time in that state's history.

Whether it's because of impatience, closer scrutiny of officeholders, or simply restlessness with the status quo, people seem to be relying more on the recall as an instrument of political change. Although the Progressive reformers intended the recall as a tool of last resort to remove those who abused the public's trust because of despicable behavior, in fact the recall has become a means of instant "political housecleaning." Along with growing numbers of lawsuits and the expanded use of negative media, it's part of the collection of activities that demands change outside of the traditional political environment. In the words of one political observer, "formerly extreme and extraordinary political and legal tactics are being used with increasing regularity. There's little reason to think that recalls will be an exception."[21]

Others aren't so sure. No less an authority than Darrell Issa said, "No, there aren't going to be more recalls. There may be a few more people trying them—to find out that you don't get a recall just because the governor is unpopular. You get it when people have given up hope that the governor can produce the outcome that they need to have."[22] Those who don't see recall as a trend view Davis's as the result of a "perfect storm" that will be replicated only under the most extraordinary circumstances. Recalls in others states are even less likely because the threshold to qualify for the ballot is higher and the standards for recall usually include specific wrongdoing, not just poor leadership or an unappealing personality. But other states aren't California, where the recall threshold remains low and professional political consultants as well as voters and interest groups will now surely view recall as one more available political weapon. California's recall election may have manifested other trends, however. Many observers thought the recall reflected anti-incumbent attitudes. "For the people to win," Arnold Schwarzenegger said repeatedly, "politics as usual must fail." But where would that stop? In California? With Congress? Or with President George Bush, a Republican? Even as the recall campaign was under way, the insurgent candidacies of outsiders like former Governor of Vermont Howard Dean and former General Wesley Clark were gaining momentum. And if a recession and budget deficit could be a problem for Gray Davis, would they be a problem for President Bush in 2004? "That same kind of anger and frustration can happen across the country if the economy doesn't improve, if the job situation doesn't improve, if gridlock in Washington continues on major issues," said Leon Panetta, a former member of Congress and chief of staff to President Clinton. "If I were an incumbent in any office, I would be a lot more nervous today."[23]

Another trend—more dangerous to democratic governance—may be hardball, take-no-prisoners partisan politics. We've seen it elsewhere: in unrelenting personal attacks on President Bill Clinton, in Florida's "battle of the chads," in redistricting in several states, in judicial confirmations in the U.S. Senate, and in many battles in the U.S. Congress. Some observers (including Gray Davis) saw the recall as part of this trend, another example of Republican opportunism. But while that may have been the motive of some, the resonance of the recall with the electorate suggests that other factors were at work, including disgust with the entrenched partisan warfare of California politics. Many voters responded affirmatively to Schwarzenegger's attacks on "politics as usual."

The Uncertainty Ahead

It remains to be seen whether the recall of Gray Davis is the result of unusual circumstances or the beginning of a new chapter in American political life. One thing is sure: Given the character issues swirling around Arnold Schwarzenegger upon his assumption of office, his honeymoon with the state may be short. The California Constitution allows recall petitions to be circulated any time after an incumbent is in office for 6 months. Whether the voters' energies have been exhausted on this issue remains a question in itself.

As to the recall process itself, here, too, there are more questions than answers. A statewide survey taken after the tumultuous recall campaign found 68 percent of the participants agreeing with a proposal to raise the number of signatures required to become a replacement candidate from the current 65

152

to 25,000. In the same survey, nearly 6 out of 10 believed that the recall process should be used only in cases of criminal or unethical wrongdoing, rather than the criteria-free basis currently in place.[24] On a related front, California state Assemblymember Mark Ridley-Thomas introduced a constitutional amendment that would change the required number of recall petition signatures from 12 percent of those who voted in the last gubernatorial election to 12 percent of the state's registered voters—in this case going from 897,158 to more than 1.8 million. Recall proponents, understandably, had a different take on the process and its accomplishments. As Rescue California recall leader Dave Gilliard said, "Some people don't like the fact that this recall qualified for the ballot. It didn't qualify because the threshold was low, but because the governor wasn't performing."[25] And the debate goes on.

12

Epilogue

While every book has an ending, *Recall!* remains a work in progress. True enough, the decision of the voters to remove Gray Davis from the governorship and replace him with Arnold Schwarzenegger closed the book on the Davis administration. Yet, the decisions made on October 7 seemed to invite countless questions on October 8 and each day after that. Would Schwarzenegger be able to transform his successful appeal from a campaign mode to a governing mode? Would the new Republican governor and the Democratic legislature be able to succeed where a Democratic governor and Democratic legislature had failed? Would the voters be satisfied with the actions of the new leadership set, or would they want more?

The first weeks of the Schwarzenegger administration provide enough clues for at least some tentative assessments. They reflect simultaneously the state's collective hunger for change, concerns for the unknown, desire for a different mode of governance, and the search for simple solutions to complex problems.

Out with the Old—In with the New

On November 17, 2003, Arnold Schwarzenegger was sworn in as California's thirty-eighth governor. His transition team had promised a simple ceremony with little public fanfare. The ceremony was simple enough, but the rest was pure Hollywood. More than 7,500 guests were invited to the capitol lawn to witness the event. As the attendees filed on to the lawn, two huge screens displayed pictures of Schwarzenegger in a variety of settings, with the photographs changing in Viewmaster-like fashion every 4 seconds. Just prior to the actual swearing in, the crowd was entertained by a children's chorus with songs from *The Sound of Music*, a conscious homage to Schwarzenegger's Austrian roots. The governor-elect arrived with his wife, Maria Shriver, and a full contingent of Shriver-Kennedy relatives. Also in attendance were movie stars Jamie Lee Curtis (*True Lies* costar), Linda Hamilton (*The Terminator* costar) and Rob Lowe.

Schwarzenegger's populist campaign notwithstanding, the swearing-in event was not open to the public; their only glimpse would be through the temporary chain-link fences that surrounded the area. After the 11:00 a.m. swearing-in, the new governor and his entourage attended three private lunches in rapid succession, including one with his allies from the California Chamber of Commerce.

The event attracted worldwide attention. All together, more than 700 members of the press received credentials. They came not only from California and the United States but from 14 other countries including Russia and Japan.[1] The British Broadcasting Corporation (BBC) attended and began collecting footage for a year-end documentary on the recall. Many California stations carried the entire event live over a period of more than 2 hours.

Schwarzenegger the Activist

The new governor wasted little time in making good on campaign promises. During his 12-minute address, he announced that he would issue an executive order repealing the increase in the car tax that had been put into place by Gray Davis immediately after the inauguration ceremony. As he penned the directive moments later, Schwarzenegger reflected on his campaign theme: "This is action, not just dialogue; this is action."[2] He also announced a freeze on all new regulations, pending comprehensive reviews from his office.

Knowing full well that some of his most important initiatives required the cooperation of others, Schwarzenegger scheduled a series of meetings with legislative leaders, union executives, and others—both former supporters and opponents—signaling his willingness to reach out. Most of those he met with were impressed by his effort and charmed by him personally. Even before taking office, he appointed a "transition team" that included Republican leaders of the California congressional delegation, state assembly, and state senate as well as Democrats such as San Francisco Mayor Willie Brown and former Assembly Speaker Bob Hertzberg.[3] While the team never met in person and may have contributed little to the actual transition, the message to the public and other politicians was one of cooperation and a different sort of politics, perhaps even an end to gridlock.

The governor also called on the legislature to work with him in a bipartisan fashion. Specifically, he asked for the repeal of SB 60, the highly unpopular new law permitting automobile driver's licenses for undocumented immigrants. He also asked the legislature to reform workers' compensation rules; although its members had done so only weeks before, the changes were not enough for a governor who wanted to restore a business-friendly environment. Most significantly, Schwarzenegger called the legislature into special session to deal with California's fiscal crisis. In the span of 12 short minutes, the new governor had acted swiftly—and had begun to deliver on his campaign promises.

Early Victories

The legislature—not known for its alacrity—responded to the governor's requests. Within 2 weeks of Schwarzenegger's proposal, legislators voted overwhelmingly to repeal SB 60, and the governor signed the bill on December 3. Original Democratic supporters of the concept retreated, intimidated by the new governor and by the very real threat of a referendum. Ironically, the author of SB 60 joined the repeal effort yet vowed to return with another bill in 2004. The repeal was facilitated by a promise from Governor Schwarzenegger to consider a compromise on a new bill that met security concerns, but the reversal confirmed the early momentum of the new administration and delivered on another campaign promise.[4] It also sent Latino leaders reeling in what could only be viewed as a bitter defeat.[5]

Schwarzenegger also succeeded with his regulatory rollbacks. The executive order put a temporary freeze on more than eighty new regulations covering areas such as cancer-causing chemicals, energy-efficient standards, water policies, and allowable levels for various carcinogenic substances. Critics accused the governor of betraying his environmental promises during the campaign. For his part, Schwarzenegger said that he wanted to see how the costs of the new regulations would affect industry competitiveness,[6] a major campaign theme. Regardless, that Schwarzenegger could act in this manner clearly demonstrated the powers of his office and his willingness to use them. It also showed supporters and foes alike that he was capable of getting things done.

Early Defeats

Along with early victories came early defeats for California's thirty-eighth governor. During the campaign, Schwarzenegger vigorously championed the need to make California business-friendly once again. This, he said, would restore the state's economic vitality. But how could such a theme be translated into public policy? Part of the answer would come by reducing the cost of workers' compensation insurance, a $15 billion program paid by employers at rates higher than those of any other state in the nation.[7] Although the legislature passed modest reform legislation in September, Schwarzenegger argued that radical change was in order. After he assumed office, the governor asked the legislature to pass "real" reform with a jobs stimulus package that included various tax incentives. The request went nowhere, at least in the first few weeks. There were other issues that remained without response. Increasing the state's take from Indian gambling revenues and renegotiating state employee contracts were high on the Schwarzenegger reform agenda after his election.[8] Both of these would-be changes languished, too, perhaps because of the all-consuming budget crisis, perhaps because of powerful forces with allies in the legislature. Whether they would be dealt with in 2004 remained to be seen.

Quick action on the budget, however, resulted in a quick lesson that every program has a constituency and every cut results in protests. Schwarzenegger proposed midyear budget cuts (subject to legislative approval) of nearly $2 billion in a variety of programs ranging from social services to higher education. Among the cuts were $274 million for programs for the developmentally disabled. Within days, parents with children in wheelchairs were demonstrating outside the Capitol. Schwarzenegger and his in-laws, the Shrivers, are well-known advocates for the developmentally disabled, however, and the embarrassed governor quickly reinstated the spending. "It was one of those things that slipped through," he said.[9]

Early Compromises

The centerpiece of Schwarzenegger's platform focused on California's continuing budget deficit. Throughout the recall campaign, he had pummeled the Davis administration and legislature for irresponsible fiscal behavior. Now it was time for "action, action, action, action," he said, challenging the legislature to address the budget issue. "That's what the people have voted me into office for."[10] But would the legislature respond?

Governor Schwarzenegger immediately appointed Donna Arduin as his director of finance. With a reputation as a tough budget manager, Arduin had worked for Florida's Republican Governor Jeb Bush. Schwarzenegger's midyear budget cuts of nearly $2 billion still left a huge budget shortfall, and the repeal of the recently increased motor vehicle fees only added to its size.

In order to address the fiscal mess comprehensively, Schwarzenegger asked the legislature to quickly approve a $15 billion bond, to make up for the current shortfall, and to cap state spending to prevent future deficits. Both proposals would appear on the March 2 ballot, and they included language that one would not move forward unless the other was also approved. According to state law, the legislature had to act on the two measures no later than December 5 to qualify them for the March 2 ballot. Before all was said and done, Secretary of State Kevin Shelley would extend the deadline twice, to December 12.

From the governor's perspective, the bond would keep the state solvent without making draconian cuts; the debt could be paid off over a period of 20 to 30 years, thereby eliminating the necessity for immediate painful decisions. As for future fiscal crises, the spending cap would prevent the state from ever again appropriating more than the available revenues.

From the earliest moments, most Democrats and a few Republicans were skittish on the bond and downright opposed to the spending cap. Regarding the bond, many Democrats fretted over the long payback period, passing current debt along to future generations. They also argued that without tax increases, the state would be faced with another deficit beginning July 1, 2004, the start of the next fiscal year. With California's credit rating ranked near the bottom among the fifty states, opponents had reason to worry. Ultimately, however, they negotiated a compromise with most Republican legislators that would pay off the bond in 9 years.

The spending cap was the major stumbling block. Schwarzenegger and most legislative Republicans insisted on a strict ceiling that would greatly constrain future spending; the proposed cap also gave the governor sweeping powers to reduce spending during the year, allowing the legislature to overturn any decisions by a two-thirds vote. Democrats countered that the cap should be more flexible, protecting public education and other social programs; they also opposed giving additional powers to a governor who already had at his disposal the "item veto"—the ability to eliminate dollar amounts from specific budget items.

Seeing major opposition in the legislature, Schwarzenegger announced that he would barnstorm the state and go to the people directly. Campaigning up and down California to capacity crowds, Schwarzenegger urged the throngs to call on their lawmakers to pass the spending cap. "The 'Never-Again Spending Limit' will be the equivalent of taking the state credit card away from the politicians," he said to cheering audiences.[11] But translating campaign rallies into legislative acts would not be so easy.

Two deadlines—first December 5, then December 10—came and went as seasoned observers waited to see who would blink first. Finally, after considerable bargaining, the governor and Democrat-controlled legislature reached a compromise at the eleventh hour. Along with the $15 billion bond, there would be the spending cap sought by Schwarzenegger, but it would be watered down to little more than a plan actually proposed by the Democrats during the Gray Davis regime, with little means to enforce budget restraints.[12] Some said the budget compromise indicated a new, bipartisan spirit in Sacramento, but others were skeptical. As veteran *Sacramento Bee* columnist Dan Walters observed, "In the jargon of political insiders, Schwarzenegger 'rolled,' and his retreat put his fellow Republicans on the spot."[13] Perhaps, but the deal was done.

The Movie Star and the Bully Pulpit

Governor Schwarzenegger's first weeks in office saw new policy proposals as well as a new political style, extending Schwarzenegger's campaign techniques to governance. The governor remained the darling of the media and the master of political symbolism, providing him with the best "bully pulpit" (access to the public) any governor has enjoyed in recent times. But could he sustain that access and could he translate it into results in Sacramento?

Before the recall election, the capitol press corps was down to about forty newspaper correspondents. For more than 20 years, no television station in the state had a Sacramento bureau except those based in the capitol itself. All that changed with the recall and the election of Governor Schwarzenegger. A hundred journalists and three dozen television cameras covered his first press conference, which was broadcast live. Television cameras became omnipresent, as satellite crews drove into Sacramento for events and news organizations from around the state and nation set up offices. Talk radio also moved to the mainstream of California politics and Governor Schwarzenegger continued to use it after his election. Interest might soon wane, but it was also clear that Governor Schwarzenegger could command the cameras and the attention of the press when he wanted to; by doing so, he had a level of access to the public unmatched by previous governors.

In his first days in office, Schwarzenegger demonstrated his mastery of that access by taking actions that sent strong messages. The appointment of the transition team made news across the state partly because the team was regionally balanced—so every area had somebody local on the team—but also because it sent a message of reaching out to differing political viewpoints and a rejection of politics as usual. Schwarzenegger followed with the appointment of several women to high-profile offices in his administration. All were fully qualified, but the appointments also helped defuse allegations about Schwarzenegger's sexism. The cordial meetings with ostensible opponents and the purportedly modest inauguration extended the appearance of rejecting politics as usual.

The governor didn't forget visuals, either. Liberals and moderates were reassured by the presence of the Shrivers and Kennedys at the inauguration. The repeal of increases in the motor vehicle registration fee (the "car tax") was announced in front of an auto dealership—a very different visual than the normal man-at-desk-with-pen-surrounded-by-smiling-legislators. And when the legislature resisted his proposals for a bond measure and spending cap, he took his show on the road with rallies. At the same time, he went back on talk radio to make the case for his reforms.

Clearly, however, not all of Schwarzenegger's actions were beyond the realm of politics as usual. He appointed many of the staffers of former governor Pete Wilson, who chaired Schwarzenegger's campaign, to important positions in his administration and he maintained a cozy relationship with the California Chamber of Commerce, appointing its lobbyist as his legislative liaison. He held a series of fund-raisers, presumably to recoup the millions of dollars he loaned his campaign. And he accepted campaign contributions from "special interests" and groups that do business with the state. In these ways, Schwarzenegger has reflected the classic "insider politics" of all elected chief executives.

But there is another side of Schwarzenegger—namely, the side that is willing to take his case directly to the voters if necessary. That's part of his strategy to move the legislature. He can do so through ballot measures, as threatened on the driver's license issue and the bond/spending cap proposals. Will the voters follow? That remains to be seen.

One fact is clear—Democratic leaders who initially offered cooperation turned resistant. Phil Angelides, the Democratic state treasurer and candidate for the 2006 Democratic nomination, quickly seized Schwarzenegger's model and went on the road to rally the public against the bond proposal, albeit with far smaller crowds. Angelides even organized statewide television commercials urging the legislature to reject the bond and spending cap proposals. Democratic legislators, angry about Schwarzenegger's unwillingness to compromise on initial budget plans and secure in their safe seats, have also shown the ability not only to resist but also to remold Schwarzenegger proposals into acceptable ideas. Even some Republican leaders who are more conservative than Schwarzenegger may also resist, as his former opponent Tom McClintock has already done on the bond proposal.

Schwarzenegger argues that his election was a mandate for change—"the people's veto of politics as usual,"[14]—but he won with 48.6 percent of the vote. Will that be enough or can he build on it? Or will he succumb to politics as usual himself, falling into the fund-raising and deal-making routines of his predecessors? While it's clear he has the capacity to reach huge audiences through the media and public appearances, can he leverage his entertainment celebrity into meaningful political clout? Will his followers abide by his exhortations to contact their legislators or even vote them out of office, or will they be appeased by autographs? Will they follow their new governor's lead on ballot measures? Will he campaign for Republican candidates for the state legislature at the risk of alienating Democrats? Can he remake California's conservative-leaning Republican Party, or will it remake him? Can a governor manage a state like California, with all its diversity, by rallying public opinion and using the tools of direct democracy? And is this approach good for running a state, anyway?

More to Come—But What?

By year's end, a political winter freeze had overcome the normally balmy climate of the Golden State. Just 2 months earlier, the state's electorate had seemed so strongly committed to changing the status quo, to correcting a foundering course, yet the precise terms of "change" remained as elusive as ever.

With the election of Arnold Schwarzenegger, one chapter has ended in the California story, but many others are yet to be penned. Yes, the governor and legislature patched together a compromise to help the state limp to the end of the fiscal year on June 30, but what after that? Would there be more taxes or steep spending cuts? Clearly, such a decision would have to be made sooner or later, notwithstanding the wishful thinking of Schwarzenegger and the legislature.

Whatever happens, the state and nation will witness some interesting moments in the months and years ahead. There is little question that a political earthquake has rocked California. Now we await the aftershocks here and elsewhere.

Appendix A

Recall Timeline

February 5, 2003: Ted Costa, head of People's Advocate, a taxpayers' group, begins campaign to recall Governor Gray Davis, reelected to office in November 2002.

February 24: Recall proponents file petition forms with the secretary of state's office.

March 25: Secretary of State Kevin Shelley authorizes recall proponents to begin collecting signatures for the recall petition of Governor Gray Davis.

April 15: The popularity of California Governor Gray Davis plunges to 24 percent in a statewide Field Poll, the lowest in the 55-year history of the poll.

May 15: Conservative Republican Congressman Darrell Issa donates $445,000 to the recall effort and announces his candidacy for governor; Issa eventually donates nearly $2 million to the recall campaign.

May 28: Davis allies launch an anti-recall campaign.

June 15: Darrell Issa officially announces his candidacy for governor.

June 17: Key Democrats Controller Steve Westly, Treasurer Phil Angelides, and Attorney General Bill Lockyer announce that they will not run as candidates; in a separate statement 2 days later, Lieutenant Governor Cruz Bustamante also declines to run.

July 9: California Republican State Chairman Duf Sundheim endorses the recall effort.

July 18: Terry McAuliffe, Chairman of the Democratic National Committee, pledges to use all resources available to defeat the recall.

July 23: Secretary of State Kevin Shelley announces that there are more than the necessary number of signatures to qualify the recall for the ballot (897,158 are necessary to qualify and 1.3 million valid signatures have been submitted).

July 24: The California State Supreme Court dismisses all arguments against the recall qualifying for the ballot.

July 25: Lieutenant Governor Cruz Bustamante sets October 7, 2003, as the date for the recall election; he says that under no circumstances will he be a candidate.

The 75-day campaign will be about half the length of traditional official campaign periods, forcing county registrars to change the ways they operate.

A total of 473 people take out filing papers.

August 3: Governor Davis's legal challenge of the recall as unconstitutional is denied. (Other suits dealing with reduced numbers of precincts and fewer interpreters were tossed out by state and federal courts.)

The suit by the ACLU alleging that "hanging chad"–producing faulty voting machines in Los Angeles County and five other counties will deny as many as 40,000 voters "equal protection under the law" (Fourteenth Amendment claim) awaits disposition by a U.S. District Court judge.

August 5: Republican State Senator Tom McClintock files for election.

August 6: Lieutenant Governor Cruz Bustamante and Insurance Commissioner John Garamendi announce intentions to run for governor.

August 6: Surprising reporters and close friends who expected otherwise, Republican actor Arnold Schwarzenegger announces on the *Tonight Show* that he will run for governor.

August 7: Former Los Angeles Mayor and Schwarzenegger friend Richard Riordan announces that he will not run as a replacement candidate for governor.

August 7: Darrell Issa decides not to enter the race.

August 9: Independent Arianna Huffington and Republican Peter Ueberroth file for election.

August 9: John Garamendi drops out, leaving Bustamante as the only top-tier Democrat.

August 9: Candidate filing ends. A total of 247 have filed papers, of whom 135 qualify.

August 20: U.S. District Court Judge Stephen Wilson dismisses the claim of the American Civil Liberties Union (ACLU) that as many as 40,000 people will be disenfranchised by faulty chad–producing punch-card machines; the ACLU appeals to the Ninth Circuit Court of Appeals.

August 22: All thirty-three Democratic members of the California House of Representatives endorse Bustamante as a replacement candidate.

August 24: Bill Simon, Republican nominee for governor in 2002, drops out.

September 8: Absentee voting begins.

September 9: Republican Peter Ueberroth drops out.

September 13: Former President Bill Clinton campaigns in California with Davis against the recall.

164

September 13: State Democratic leaders formerly adopt the "No on Recall, Yes on Bustamante" strategy; Bustamante and Davis appear briefly together.

September 13: State Republican leaders meet in Los Angeles. McClintock challenges Schwarzenegger to debate; Schwarzenegger declines.

September 14 to 15: Former President Bill Clinton campaigns for Gray Davis.

September 15: A three-judge panel of the Ninth U.S. Circuit Court of Appeals overturns a lower court ruling and declares that the use of old chad-producing voting machines in six counties with 44 percent of the state's voters violates the "equal protection" clause of the Fourteenth Amendment. The court reschedules the election for March 2, 2004.

September 16: The full Ninth Circuit Court of Appeals asks both sides to submit arguments on their positions by September 17.

September 19: The Ninth Circuit Court of Appeals orders oral arguments on the case to take place on Monday, September 22. Per court procedures, the expanded en banc panel consists of the chief justice and ten other justices of the Ninth Circuit Court selected at random.

September 22: A Sacramento judge finds that $3.8 million in campaign contributions to Lieutenant Governor Cruz Bustamante by Indian casino interests violated Proposition 34 spending limits; he orders Bustamante to stop using the money, but campaign spokespersons say that it has been spent.

September 22: Both sides of the issue speak before the Ninth Circuit Court of Appeals for 30 minutes; eight of the eleven judges were appointed by Democrats, but the group is considered more moderate than the three-judge panel.

September 22: The last day for anyone to register to vote.

September 23: The Ninth Circuit Court of Appeals overrules the three-judge panel by an 11-to-0 vote, restoring the election date of October 7. The ACLU, plaintiff in the case, decides not to appeal the case to the U.S. Supreme Court.

September 23: Conservative Republican State Senate Minority Leader Jim Brulte, chairman of McClintock's 2002 campaign for state controller, endorses Arnold Schwarzenegger.

September 24: The five remaining "major" candidates (Bustamante, Schwarzenegger, McClintock, Camejo, and Huffington) debate for the first—and only—time during the campaign.

September 25: Republican county chairpersons meet in Sacramento in an attempt to pry McClintock from the campaign; he refuses, and forty-two of fifty-six county chairpersons endorse Schwarzenegger.

September 25: Bill Simon, Republican gubernatorial candidate in 2002 and replacement dropout candidate in 2003, endorses Schwarzenegger.

September 25: Democratic Governor Gray Davis challenges Arnold Schwarzenegger to a one-on-one debate; Schwarzenegger declines.

September 26: Darrell Issa, the Republican Congressman who financed the recall effort with $2 million of his own money, endorses Schwarzenegger.

September 29: The board of directors of the California Republican Party endorses Schwarzenegger.

September 30: Arianna Huffington drops out, offering the need to stop Arnold Schwarzenegger as the most important reason.

October 1: A *Los Angeles Times* poll shows Gray Davis losing the recall by 14 points and Arnold Schwarzenegger defeating Cruz Bustamante in the replacement election by 8 points.

October 2: The *Los Angeles Times* publishes a story that has six women accusing Arnold Schwarzenegger of making improper sexual advances over a period of nearly 30 years. He apologizes to those he offended but lashes out at the newspaper for "trash politics" at the eleventh hour of the campaign.

October 2: In what turns out to be one of the most bizarre moments of the campaign, Arianna Huffington—a replacement candidate until 2 days earlier—speaks at a debate on behalf of rejecting the recall and keeping Gray Davis in office.

October 4: The *Oakland Tribune* retracts its endorsement of Arnold Schwarzenegger as its replacement choice for governor.

October 5: The *Los Angeles Times* reports that the number of women claiming improper sexual advances by Arnold Schwarzenegger has reached fifteen; within a week, more than 1,000 customers cancel their subscriptions in protest of what Schwarzenegger calls "trash politics."

October 7: Election day; Davis is recalled and Schwarzenegger is elected governor.

November 14: Secretary of State Kevin Shelley officially certifies that Arnold Schwarzenegger has won the replacement contest for governor.

November 17: Arnold Schwarzenegger is sworn in as California's thirty-eighth governor.

Appendix B

The 2003 Recall Election and California Politics on the Internet

California's Home Page: www.ca.gov

Office of the Governor: www.governor.ca.gov

California Elections Code
- California Law on Recall, etc.: www.leginfo.ca.gov/cgi-bin/
  calawquery?codesection=elec&codebody=&hits=20

California Secretary of State (www.ss.ca.gov)
- Voter Registration Statistics: www.ss.ca.gov/elections/elections_u.htm
- California Recall History: www.ss.ca.gov/elections/sov/2003_special/
  contests.pdf
- Proponent's Grounds for Recall and The Governor's Response:
  www.ss.ca.gov/elections/recall_notice.pdf or www.voterguide. ss.ca.gov/
  recall/recall.html
- Map of Recall Vote by County: vote2003.ss.ca.gov/Returns/recall/
  mapN4.htm
- Vote on Candidates to Succeed Gray Davis: vote2003.ss.ca.gov/Returns/
  gov/00.htm
- Map of Winners (Schwarzenegger or Bustamante) by County:
  vote2003.ss.ca.gov/Returns/gov/mapB.htm

Elections and Ballot Measures
- California Secretary of State: www.ss.ca.gov
- California Voter Foundation: www.calvoter.org/recall
- League of Women Voters: www.ca.lwv.org
- Smart Voter: www.smartvoter.org

Campaign Finance
- California Secretary of State: www.ss.ca.gov
- California Common Cause, www.recallmoneywatch.com

Public Opinion Polls
- The Field Poll: www.field.com/fieldpoll
- Public Policy Institute of California Statewide Surveys: www.ppic.org
- *Los Angeles Times* Poll: www.latimes.com/timespoll

News Media
- Rough and Tumble (daily summaries of news articles about California politics): www.rtumble.com
- *Los Angeles Times*: www.latimes.com
- *New York Times*: www.nytimes.com
- *San Francisco Chronicle*: (free access to archives): www.sfgate.com
- *San Jose Mercury News*: www.bayarea.com
- Grade the News at Stanford University: www.gradethenews.org

Recall Campaigns/Committees
- Davis Recall.com Committee (People's Advocate): www.davisrecall.com
- Recall Gray Davis Committee (Kaloogian/Russo): www.recallgraydavis.com
- Rescue California (Darrell Issa's Group): www.rescuecalifornia.com
- Californians Against the Costly Recall: www.no-recall.com
- Meetup.com: www.notocaliforniarecall.meetup.com

Candidates and Campaigns
- Cruz Bustamante: www.noonrecallyesonbustamante.com or www.ltg.ca.gov
- Gray Davis: www.no-recall.com
- Arianna Huffington: www.votearianna.com
- Tom McClintock: www.tommcclintock.com
- Arnold Schwarzenegger: www.joinarnold.com
- Campaign consultants discuss tactics in "The Race to Recall—The View from the Finish Line: The Berkeley Post Mortem," available as video-on-demand, www.uctv.tv

Interest Groups
- California Chamber of Commerce: www.calchamber.com
- California Common Cause: www.commoncause.org/states/california
- California Labor Federation: www.calaborfed.org
- Howard Jarvis Taxpayers Association: www.hjta.org
- The People's Advocate: www.peoplesadvocate.org

Notes

Notes to Chapter 1

1. See Thomas E. Cronin, *Direct Democracy* (Cambridge, MA: Harvard University Press, 1989), for an excellent overview of direct democracy in America.
2. Quoted in George E. Mowry, *The California Progressives* (Berkeley, CA: University of California Press, 1951), p. 139.
3. For two very different accounts of the initiative, see Elisabeth R. Gerber, Arthur Lupia, Matthew D. McCubbins, and D. Roderick Kiewiet, *Stealing the Initiative* (Upper Saddle River, NJ: Prentice Hall, 2001); and Jim Shultz, *The Initiative Cookbook* (San Francisco: The Democracy Center, 1998).
4. California Secretary of State.
5. Alan Greenblatt, "Total Recall," *Governing* (September 2003): 25–27.
6. Department of Defense, *Prime Contract Awards by State, Fiscal Year 1981*; and Joseph Wahed, "Good Growth Ahead for California Aerospace-Electronics," *Business Review, Wells Fargo Bank* (September-October 1981): 1–2.
7. California Employment Development Department, June 2002, www.edd.ca.gov; and U.S. Department of Commerce, Bureau of Economic Analysis, *Survey of Current Business*, October 2002.
8. "The 'Silver' Age of State's Defense-Aerospace Economy," *Los Angeles Times*, July 7, 1996.
9. "Valley of the Grim," *San Jose Mercury News*, March 1, 2003.
10. Public Policy Institute of California (www.ppic.org), Statewide Survey, August 2003.
11. U.S. Census, 2000.

Notes to Chapter 2

1. Sarah McCally Morehouse and Malcolm E. Jewell, *State Politics, Parties and Policy*, 2nd ed. (Lanham, MD: Rowman & Littlefield, 2003), p. 151.
2. As of 1996, three companies—Southern California Edison, San Diego Gas and Electric, and Pacific Gas and Electric—sold power to 24 million of the state's 32 million residents.
3. See "How State's Consumers Lost with Electricity Deregulation," *Los Angeles Times*, December 9, 2000.
4. "California's Power System: What Went Wrong," *San Jose Mercury News*, January 7, 2001.

5. "Davis Faces Growing Criticism on Power Crisis," *Los Angeles Times*, December 31, 2000.

6. Comments by former Assembly Majority Leader Fred Keeley, Santa Clara County Manufacturing Group, October 3, 2003.

7. "Power Suppliers Accused of Manipulating Prices," *Los Angeles Times*, November 23, 2000.

8. See "Juice Squeeze: As California Starved for Energy, U.S. Businesses Had a Feast" *Wall Street Journal*, September 16, 2002; and "Judge Concludes Energy Company Drove Up Prices," *New York Times*, September 24, 2002. Information even more damaging to the credibility of energy companies and the Federal Energy Regulatory Commission emerged in 2003 in documents provided by the federal agency. See "Energy Probe Turns Heat on Reliant," *San Francisco Chronicle*, April 20, 2003.

9. "Tall Tales Blur Truth Behind Power Drama," *San Jose Mercury News*, March 4, 2001.

10. See "Power Deals Exceed Prices on Spot Market," *Los Angeles Times*, June 6, 2001.

11. "Californians and Their Government," *PPIC Statewide Survey* (San Francisco: Public Policy Institute of California, December 2001), p. 18.

12. "Californians and Their Government," *PPIC Statewide Survey* (San Francisco: Public Policy Institute of California, September 2002), p. 5.

13. "Actual Lost Jobs Decline Despite State's Increase in Mass Layoffs," *Birmingham Business Journal*, Birmingham, Alabama, week of April 28, 2003.

14. *Finance Bulletin*, State of California, August 2003, p. 1.

15. "Growth in Tech Jobs in California Fell 90% in 2001, Study Finds," *San Francisco Chronicle*, June 26, 2003.

16. "Tech Jobs Become State's Unwanted Big Export," *Los Angeles Times*, December 12, 2002.

17. "Report: Calif. Economy Should Pick up in 2004," *San Jose Mercury News*, September 24, 2003.

18. "Poll: Voters Negative about State Economy," *San Jose Mercury News*, August 27, 2003.

19. "State's Economy Gives Budget a Huge Boost," *San Jose Mercury News*, January 15, 2000.

20. The data here and in the following two paragraphs come from "Huge State Budget Gap Rooted in Three Major Spending Areas," *Los Angeles Times*, September 28, 2003.

21. "Tough Times Predicted for State Budget," *Los Angeles Times*, October 9, 2001.

22. These data are taken from "Davis Unveils Budget Plan Based on Cuts, Optimism," *Los Angeles Times*, January 11, 2002.

23. See Carl E. Van Horn, ed., *The State of the States*, 3rd ed. (Washington, D.C.: CQ Press, 1996), pp. 84–94.

Notes to Chapter 3

1. For a discussion of the diminishing status of the modern state legislature vis à vis the executive branch, see Alan Rosenthal, *The Decline of Representative Democracy* (Washington, D.C.: CQ Press, 1998), Chapter 8.

2. A typical example occurred in 2001, the last of the "easy" budget years, when majority Democrats pried loose the last three Republican votes by providing the hold-

outs with $6 million for the Lodi and Galt police departments, special tax breaks for certain forestry and farming businesses, and aid for farmers in the economically depressed Klamath Valley. See "Assembly Approves State Budget," *Los Angeles Times*, July 17, 2001; and "Crude Budget Bribery," *Los Angeles Times*, July 18, 2001.

3. "Behind California Budget Mess: A Pattern of Political Paralysis," *Wall Street Journal*, January 10, 2003.

4. "GOP Budget Plan Axes $25 Billion," *San Francisco Chronicle*, February 27, 2003.

5. "Careers at Stake, Brulte Tells GOP," *Los Angeles Times*, June 5, 2003.

6. "Voters Very Dissatisfied with State Budget Negotiations," The Field Poll, Release #2074, July 15, 2003.

7. "Very Different Voter Reactions to Initiative to Lower Budget Approval to 55% Depending on How Measure Is Couched," The Field Poll, Release #2094, September 18, 2003. When asked directly about lowering the budget threshold to 55 percent, survey respondents rejected the proposal by a margin of 44 to 38 percent.

8. Quoted in "A Budget Process Built to Fail," *Los Angeles Times*, July 29, 2003.

9. Quoted in Kevin O'Leary, "Time's Up: Under Term Limits, California's Legislative Engine Sputters," *The American Prospect*, December 17, 2001, p. 33.

10. "Who Are These People Called Lobbyists?" *California Journal*, March 2002, p. 25.

11. The *Baker v. Carr* decision established the principle of "one man, one vote," or legislative districts of equal sizes, in the name of the Fourteenth Amendment's "Equal protection under the law" clause.

12. "Plan to Redraw District Passes," *Los Angeles Times*, September 14, 2001.

13. For a discussion on resolution of the redistricting stalemate of 1991, see Larry N. Gerston and Terry Christensen, *California Politics and Government: A Practical Approach*, 2nd ed. (Belmont, CA: Wadsworth, 1993), pp. 46–47.

14. "*California Journal*'s Election Analysis," *California Journal* (October 2002): 20.

Notes to Chapter 4

1. From the governor's official biography at www.smartvoter.org/2002.

2. "Governor Seeks to Show Lighter Side," *San Jose Mercury News*, August 31, 2003.

3. Ibid.

4. "Feinstein Star of TV Ads to Defend Davis," *San Jose Mercury News*, September 3, 2003.

5. Michael Lewis, "All Politics Are Loco!!!" *New York Times*, September 28, 2003.

6. Both the Brown and Eastin quotes are from "Governor Seeks to Show Lighter Side," op cit.

7. Perhaps the best example of this awkward relationship occurred in 1979 when, with Democratic Governor Jerry Brown campaigning for the presidential nomination out of state, Republican Lieutenant Governor-turned Acting Governor Mike Curb attempted to appoint a judge and sign a proclamation weakening the state's antismog laws. Brown raced back to the state to reverse Curb's efforts, nevertheless leaving people scratching their heads about "who's in charge." See Terry Christensen and Larry N. Gerston, *Politics in the Golden State* (Boston: Little, Brown), p. 106.

8. "Tensions Flare Between Davis and His Democrats," *Los Angeles Times*, July 22, 1999.

9. "Davis Demands Judges Reflect His Stances," *San Jose Mercury News*, March 1, 2000.

10. "Davis Attitude Brews Challenges," *San Jose Mercury News*, March 25, 2000.

11. Larry N. Gerston and Terry Christensen, *California Politics and Government*, 4th ed. (Fort Worth, TX: Harcourt Brace, 1997), p. 86.

12. "Tribes Come of Age," *California Journal* (October 1999): 13–14.

13. "Tribes Trash Davis Over Gaming Plan," *San Francisco Chronicle*, January 23, 2003.

14. "Money Flowing Past Loopholes in Recall Race," *New York Times*, August 30, 2003.

15. Larry N. Gerston and Terry Christensen, *California Politics and Government*, 7th ed. (Belmont, CA: Wadsworth, 2003).

16. Dan Walters, "Davis Again Needs Help from Liberals He Once Shunned," *Sacramento Bee*, July 21, 2003.

17. Richard J. Riordan, "Set the Voters Free," *New York Times*, October 31, 2003.

18. Los Angeles Times Poll, www.latimes.com/timespoll.

19. See "The Davis Record," *California Journal* (September 2003): 30, for a summary of the governor's record.

20. See Daniel Borenstein, "Gray Memories," *California Journal* (November 2003).

21. Dan Walters, "Doing the 'Smart' Political Thing Just Kept Backfiring on Davis," *Sacramento Bee*, October 10, 2003.

22. "California's Tax Myth," *San Francisco Chronicle*, August 31, 2003.

23. "Veteran Politician Changes Approach," *San Jose Mercury News*, October 5, 2003.

Notes to Chapter 5

1. "Kaloogian: Recall Was Good for Golden State," *North County Times*, October 18, 2003.

2. "California Recall," KPCC Radio, November 12, 2003.

3. "Kern Politico Helped Spark Recall Furor," *Bakersfield Californian*, September 21, 2003; and "California Recall," op cit.

4. "The Recall: How We Got Here," *San Jose Mercury News*, July 24, 2003.

5. www.peoplesadvocate.org/ted.html

6. Terry Christensen and Larry N. Gerston, *The California Connection*, 2d ed. (Glenview, IL: Scott, Foresman, 1988), p. 169.

7. "A Contrarian with a Mission," *Sacramento Bee*, July 24, 2003.

8. "Recall Puts Anti-tax Rebel in Spotlight," *San Jose Mercury News*, July 4, 2003.

9. Bill Schneider, "Modern Revolutionary Behind Davis Defeat," October 10, 2003, www.cnn.com/2003/ALLPOLITICS/10/10/ip.pol.opinion.costa.

10. "Kaloogian," op cit.

11. "'Nutballs' No More, Talk Radio Jocks Bask in Their Recall Role," *San Francisco Chronicle*, October 9, 2003.

12. "The Rise of the Voters," *California Journal* (September 2003): 25.

13. Ibid.

14. Michael Lewis, "All Politics Are Loco!" *New York Times*, September 28, 2003.

15. "'Nutballs' No More," op cit.

16. Ibid.

17. "Kern Politico Helped Spark Recall Fervor," *Bakersfield Californian*, September 21, 2003.

18. "Recall Proponent Comes to Town," *North County Times*, June 18, 2003.

19. Ibid.

20. "The Recall," op cit.

21. "Foes of Davis Recall Post Tactics on Web; Petition Workers Complain of Intimidation," *San Francisco Chronicle*, June 27, 2003.

22. "State Democrats Recall Lessons of Florida, *San Francisco Chronicle*, June 30, 2003.

23. Dana Wilkie, "Darrell Issa," *California Journal* (August 2003): 29.

24. "State GOP Strongly Endorses Recall," *San Francisco Chronicle*, July 10, 2003.

25. Mark Sandalow, "Bush and the Recall," *California Journal* (September 2003): 31.

26. "State GOP Strongly Endorses Recall," op cit.

27. "Bush Plans Appearance with Schwarzenegger," *New York Times*, October 9, 2003.

28. Sandalow, "Bush and the Recall," op cit.

29. *Esquire*, July 2003.

30. "Percent Signing Recall Petitions," *San Jose Mercury News*, July 27, 2003.

31. "'Perfect Storm' of Events Converged for Seeds of Recall to Take Root," *Sacramento Bee*, October 8, 2003.

32. "A Contrarian on a Mission," *Sacramento Bee*, July 24, 2003.

Notes to Chapter 6

1. "Voters Show Uncommon Interest in Recall; 58% Want Governor Removed from Office in New State Poll," *San Francisco Chronicle*, August 21, 2003.

2. Quoted in "Davis Recall Drive Hurting for Cash," *San Francisco Chronicle*, April 27, 2003.

3. "At a Dollar Per Signature, Recall Effort Is a Living," *Los Angeles Times*, July 7, 2003.

4. "Davis Reelection Team Regroups to Fight Recall," *Los Angeles Times*, June 2, 2003.

5. "Davis Seeks to Shift Focus to Recall Backers," *Los Angeles Times*, June 26, 2003.

6. "Contentious Davis Blasts GOP 'Power Grab,'" *San Francisco Chronicle*, August 20, 2003.

7. "Davis Steers Clear of Report of His Opponent's Groping," *San Francisco Chronicle*, October 3, 2003.

8. "Davis Seizes on Reports About Schwarzenegger," *Los Angeles Times*, October 4, 2003.

9. "Davis Is Battling Image of Aloofness," *Los Angeles Times*, October 5, 2003.

10. Quoted in "Schwarzenegger In, Feinstein Out," *Los Angeles Times*, August 7, 2003.

11. Quoted in "Riordan 'Stunned' by Friend, Aide Says," *Los Angeles Times,* August 7, 2003.

12. "Ringmaster at His Own Media Circus," *San Francisco Chronicle*, August 7, 2003.

13. "Visa, License Issues Pursue Candidate," *San Jose Mercury News*, September 17, 2003.

14. "Questions on NAACP Boss' Racism Charge," *San Francisco Chronicle*, October 3, 2003.

15. "Actor Has Poor Voting Record," *San Francisco Chronicle*, August 12, 2003.

16. Ironically, during the *Oprah Winfrey Show*, at the precise moment that Schwarzenegger was attempting to deflate accusations of sexism, he made a remark that caused his own wife to put her hand over his mouth to stop further discussion. See "Schwarzenegger's Words About Women Are at Issue," *Los Angeles Times*, September 16, 2003.

17. "Recall Madness: An Experiment in Democracy, and Satire," *Los Angeles Times*, October 6, 2003.

18. Quoted in "Victory for Arnold," *U.S. News and World Report*, October 20, 2003.

19. This theme, first expressed on the *Tonight Show*, was stated throughout the Schwarzenegger campaign. See "Schwarzenegger Steals Recall Scene," *San Francisco Chronicle*, August 7, 2003.

20. Quoted in "Schwarzenneger's Liberal Views Leave GOP Flummoxed," *San Francisco Chronicle,* August 14, 2003.

21. "Schwarzenegger Takes His Campaign to the State Fair," *Los Angeles Times*, September 2, 2003.

22. Quoted in "Schwarzenegger Calls for Deep Cuts," *Los Angeles Times*, August 21, 2003.

23. "Bustamante Proposes Increases in Taxes, Fees," *San Jose Mercury News*, August 26, 2003.

24. "Bustamante Gains Labor Endorsement," *San Jose Mercury News*, August 27, 2003.

25. Quoted in "Chance Has Looked Kindly on California's No. 2 Official," *New York Times*, August 22, 2003.

26. "Court Ruling Forces Bustamante to Return Campaign Donations," *San Jose Mercury News*, September 22, 2003.

27. "Lack of Stage Presence Undercuts Bustamante," *Los Angeles Times*, October 3, 2003.

28. See Larry N. Gerston and Terry Christensen, *California Politics and Government: A Practical Approach*, 4th ed. (Fort Worth, TX: Harcourt Brace, 1997), p. 53.

29. "It's Not Just the Fringes That Keep McClintock in the Fray," *Los Angeles Times*, October 2, 2003.

30. "Verbal Jousting: Issues Play Second Fiddle to Jabs, Taunts, Barbed One-Liners," *Los Angeles Times*, September 26, 2003.

31. "Peter Camejo: He Unveils a Plan to Hike Taxes on State's Richest, Cut Levies for Others," *Los Angeles Times*, August 29, 2003.

32. Quoted in "Simon Exits Race, Citing a Crowded GOP Field," *Los Angeles Times*, August 24, 2003.

33. Quoted in "Ueberroth Sees Himself as a Problem-Solver," *Los Angeles Times*, August 16, 2003.

34. "Candidate Quits California Recall Race," *New York Times*, October 1, 2003.

Notes to Chapter 7

1. Gary Jacobson, "Partisanship in California Politics," paper delivered at the Annual Meeting of the Western Political Science Association, March 22–24, 2002.

2. *Los Angeles Times* Poll, Study #487, *Los Angeles Times*, September 12, 2003. (The issues ranked the same in the Times Poll at the end of September.)

3. PPIC Statewide Surveys, August and September 2003, www.ppic.org.

4. "McClintock Drives Anger on Car Tax," *San Jose Mercury News*, September 19, 2003.

5. PPIC Statewide Survey, June 2003, www.ppic.org.

6. PPIC Statewide Survey, September 2003, www.ppic.org.

7. "California's Tax Myth," *San Francisco Chronicle*, August 31, 2003.

8. "Schwarzenegger Talks Tough, Describes Approach," Associated Press, August 20, 2003.

9. "Schwarzenegger Adds Details to Business-friendly Policies," *San Jose Mercury News*, September 6, 2003.

10. PPIC Statewide Survey, September 2003, www.ppic.org.

11. "Your Taxes: How California Compares," *San Jose Mercury News*, September 21, 2003.

12. "California's Tax Myth," op cit.

13. "Schwarzenegger Tackles Education Policy," *San Jose Mercury News*, September 11, 2003.

14. "Discussing the Issues," *San Jose Mercury News*, September 28, 2003.

15. Greg Palast, "Arnold Unplugged," October 3, 2003, www.gregpalast.com. Enron memos on the meeting were posted on the Internet by the Foundation for Taxpayer and Consumer Rights, www.ConsumerWatchdog.org.

16. The Field Poll, Release #2039, September 10, 2003.

17. "In California, Davis and Schwarzenegger Split the Pronunciation Vote," *New York Times*, September 9, 2003.

18. "State Senate Calls on Davis to Apologize for Accent Remarks," *San Jose Mercury News*, September 10, 2003.

19. "Alleged Sexual Harassment Confronts Schwarzenegger in Final Days," *San Jose Mercury News*, October 3, 2003.

20. "As Actor Lashes Out at Allegations Demos Fire Back on Several Fronts," *San Jose Mercury News*, October 5, 2003.

21. "Huffington Campaigns on Need to Remake the State," *San Jose Mercury News*, September 22, 2003.

22. "Schwarzenegger Woos Corporate Donors," *San Francisco Chronicle*, September 7, 2003.

23. "Seizing the Moment and Defying Expectations," *New York Times*, October 9, 2003.

24. *Los Angeles Times* Poll, Study #487, September 12, 2003.

25. "Reality Check," *San Jose Mercury News,* September 4, 2003.

Notes to Chapter 8

1. Comments on "The California Recall," KPCC Radio, November 12, 2003.

2. "The Race to Recall—The View from the Finish Line: The Berkeley Post Mortem," available as video-on-demand, www.uctv.tv.

3. "Bustamante Defends Student Background. Conservatives See Racism in Separatist Group," *San Francisco Chronicle*, August 29, 2003.

4. See Kevin Starr, *Endangered Dreams* (New York: Oxford, 1996), for a powerful history of labor activism in California in the 1930s.

5. "Recall: Huffington Out; Actor Surges in Poll," *San Jose Mercury News*, October 1, 2003.

6. From a study by the Rockefeller Institute of Government, cited in "California's Tax Myth," *San Francisco Chronicle*, August 31, 2003.

7. "Schwarzenegger Wooed Outsiders, Insiders Alike," *Sacramento Bee*, October 8, 2003.

8. "Giving Millions to Back a Winner," *Sacramento Bee*, October 12, 2003.

9. Ibid.

10. Ibid.

11. "Business Lobbyist Tapped as Aide," *San Jose Mercury News*, November 11, 2003.

12. "Interest Groups Played Key Recall Role," Associated Press, October 15, 2003.

Notes to Chapter 9

1. "Recall Race Starts to Look Conventional," *New York Times*, September 30, 2003.

2. "TV's Intense Glare Makes the Odd California Campaign Seem Even Odder," *New York Times*, October 2, 2003.

3. "Schwarzenegger Slights Democrat-Heavy Bay Area," *San Jose Mercury News*, November 27, 2003.

4. "Actor's Staff Offers Packet Targeting Bustamante," *San Jose Mercury News*, September 5, 2003.

5. "Politicians Turn to Focus Groups to Plot Strategy," *San Francisco Chronicle*, August 25, 2003.

6. "The Race to Recall—The View from the Finish Line: The Berkeley Post Mortem," available as video-on-demand, www.uctv.tv.

7. See polling data at www.field.com/fieldpoll/, www.latimes.com, and www.ppic.org.

8. "The Race to Recall," op cit.

9. On the complexity of polling in the recall election, see "Pollsters Groping for Questions," *Los Angeles Times*, August 13, 2003; "Why Poll Results Differ," *Los Angeles Times*, September 12, 2003; and "A Different Take on 'Why Polls Differ,'" The Field Poll, September 16, 2003.

10. PPIC, Statewide Survey, September 2003, www.ppic.org.

11. "Absent Actor Is Targeted at Lively Candidates Debate," *San Jose Mercury News*, September 18, 2003.

12. "Schwarzenegger Approach Would Be Hard to Copy," *Los Angeles Times*, October 10, 2003.

13. "Absent Actor," op cit.

14. "Actor Shows He's Credible as Governor," *San Jose Mercury News*, September 25, 2003.

15. "Actor's Drive, Public's Ire Clear Road to Sacramento," *San Diego Union Tribune*, October 8, 2003.

16. "The Race to Recall," op cit.

17. Ibid.

18. "Seizing the Moment, and Defying Expectations," *New York Times*, October 9, 2003.

19. "Taking Charge," *San Jose Mercury News*, October 5, 2003.

20. "My Economics," *Wall Street Journal*, September 24, 2003.

21. "Celebrity Status Gave Schwarzenegger Options," *San Francisco Chronicle*, October 13, 2003.

22. "Republican Front-Runner Acts as If Win Is Inevitable," *San Jose Mercury News*, October 2, 2003.

23. Michael Stoll, "Papers Couldn't Resist Schwarzenegger," www.gradethenews. org.

24. "'Perfect Storm' of Events Converged for Seeds of the Recall to Take Root," *Sacramento Bee*, October 8, 2003.

25. "Governor Candidates Bask in Media Coverage," *San Jose Mercury News*, September 7, 2003.

26. Comments of both Garry South and the editor of the *Los Angeles Times* are from "Race to Recall," op cit.

27. Ibid.

28. PPIC, Statewide Survey, September 2003, www.ppic.org.

29. "California Recall," KPCC Radio, November 12, 2003.

30. "Celebrity Status Gave Schwarzenegger Options," op cit.

31. *Late Show with David Letterman*, August 7, 2003.

32. For an extended discussion of celebrity politics, see Darrel M. West and John Orman, *Celebrity Politics* (Upper Saddle River, NJ: Prentice Hall, 2003).

33. "Tabloids Starry-eyed for Schwarzenegger," *San Jose Mercury News*, September 26, 2003; and "Schwarzenegger Prompts Role Reversal Among Media," *New York Times*, October 6, 2003.

34. "World Press Digests Arnie's Latest Role," BBC News, October 21, 2003; "Election (and Arnold) Captivate the World's Press," *Christian Science Monitor*, August 25, 2003; "World Marvels at Schwarzenegger," MSNBC News, October 8, 2003; "Outsiders Agog at Pick of Voters," *San Francisco Chronicle*, October 9, 2003.

Notes to Chapter 10

1. Quoted in "Cost of Davis Recall Election: $25 million; State Could Have to Pay His Costs If He's Victorious," *San Francisco Chronicle*, February 18, 2003.

2. See "Voters Very Dissatisfied with State Budget Negotiations," The Field Poll, Release #2074, July 15, 2003.

3. "Davis, Fighting Recall, Is Ready to Stump Against 'Right Wing,'" *New York Times*, July 26, 2003.

4. Quoted in "Davis Goes Back to His Base," *Los Angeles Times*, May 20, 2003.

5. *Los Angeles Times* Poll Alert, August 24, 2003.

6. "Bustamante Leads Schwarzenegger in Recall Race," *Los Angeles Times*, August 23, 2003.

7. Interview with NBC11 reporter Beth Willon, Los Angeles, California, November 14, 2003.

8. "License Bill Draws Mixed Response," *Los Angeles Times*, July 29, 2003.

9. "California Governor Attempts Belated Personality Makeover," *New York Times*, September 1, 2003.

10. "The Recall Election: A Losing Effort," *Los Angeles Times*, October 8, 2003.

11. "Wanted: A Look," *Los Angeles Times*, August 19, 2003.

12. "Increased Majority of Voters Ready to Recall Davis," The Field Poll, Release #2081, August 15, 2003.

13. Interview with NBC11 reporter Beth Willon, Los Angeles, California, November 14, 2003.

14. "Davis, the Great Un-communicator," *San Francisco Chronicle*, October 8, 2003.

15. "Voters in a State of Change," *Los Angeles Times*, October 8, 2003.

16. "Davis Seeks a Debate; Schwarzenegger Camp Says No," *New York Times*, September 27, 2003.

17. "Davis Is Told: No Trash Talk," *Sacramento Bee*, August 1, 2003. Democrat Lockyer himself became somewhat controversial after the election when he announced that while he had voted "no" on the recall, he had voted for Republican Schwarzenegger as his replacement choice for governor. See "Lockyer's Shocking Choice in Recall," *San Francisco Chronicle*, October 19, 2003.

18. "Davis Steers Clear of Report of His Opponent's Groping," *San Francisco Chronicle*, October 3, 2003.

19. "Davis Seizes on Reports About Schwarzenegger," *Los Angeles Times*, October 4, 2003.

20. California Recall Election Exit Poll, conducted for NBC News by Edison Media Research/Mitofsky International, October 7, 2003.

21. Data here and in the following paragraphs of the section come from an exit poll appearing in the *Los Angeles Times*, October 8, 2003.

22. "In California, Davis Sues to Avert Recall 'Train Wreck,'" *New York Times*, August 5, 2003.

23. "Delay Could Allow Recall Fever Time to Cool," *New York Times*, September 16, 2003.

24. "Authorities on Voting Dispute Punch-card Issue," *San Francisco Chronicle*, September 17, 2003.

25. "Few Problems Reported at Polling Places," *San Jose Mercury News*, October 8, 2003.

Notes to Chapter 11

1. "State Budget Briefing Is Grim," *Los Angeles Times*, October 24, 2003.

2. Quoted in "Late-Night Comic," *San Francisco Chronicle*, October 11, 2003.

3. "Voters Very Dissatisfied with State Budget Negotiations; Blame Davis and Both Parties in the Legislature," The Field Poll, Release #2074, July 15, 2003.

4. "Economy Worries Californians, But It's Not That Bad," *New York Times*, October 12, 2003.

5. "The Recall Election: Exit Poll," *San Jose Mercury News*, October 8, 2003.

6. "Schwarzenegger Met the Press His Way," *San Francisco Chronicle*, October 13, 2003.

7. Quoted in "Schwarzenegger Met the Press His Way," op cit.

8. "How Davis's Lack of leadership Led to His Early Downfall," *Los Angeles Times*, October 20, 2003.

9. "Schwarzenegger Calls for Deep Cuts," *Los Angeles Times*, August 21, 2003.

10. "Hijinks: While the Cat's Away," *California Journal* (November 2003): 7.

11. "On State Tax Policy, Everyone Has a Formula for Reform," *Los Angeles Times*, February 25, 2003.

12. "Davis Takes Heat from All Sides," *Los Angeles Times*, April 13, 2003.

13. Kathleen Les, "Women Unhappy with New District Patterns," *California Journal* (December 2001): 35.

14. "Recall Mastermind Focuses on Redistricting Reform," *San Francisco Chronicle*, November 1, 2003.

15. For a comprehensive discussion on the value of endorsements and state-to-state differences, see Malcolm E. Jewell and Sarah M. Morehouse, *Political Parties and Elections in American States*, 4th ed. (Washington, D.C.: CQ Press, 2002), pp. 106–114.

16. "Lockyer Broke Ranks, Voted GOP," *Los Angeles Times*, October 19, 2003.

17. For an excellent review of the multifaceted roles of political parties, see John Kenneth White and Daniel M. Shea, *New Party Politics* (Boston: Bedford/St. Martin's, 2000).

18. "For GOP, Glow of Victory May Hide Rifts Only Briefly," *San Jose Mercury News*, October 9, 2003.

19. Quoted in "How Davis's Lack of Leadership Led to His Early Downfall," *Los Angeles Times*, October 20, 2003.

20. Quoted in "Recall Bid Could Launch Another National Trend," *Los Angeles Times*, August 12, 2003.

21. Alan Greenblatt, "Total Recall," *Governing* (September 2003): 27.

22. Darrell Issa speaking to the Commonwealth Club of California, September 22, 2003.

23. "Aftershocks Are Unpredictable," *Washington Post*, October 8, 2003.

24. "Most Voters Support Changing Recall Steps," *San Francisco Chronicle*, October 16, 2003.

25. "Bill Is Proposed to Revise California's Recall Process," *New York Times*, October 22, 2003.

Notes to Chapter 12

1. "A State Event Covered by the World's Media," *Los Angeles Times*, November 18, 2003.

2. "Schwarzenegger Sworn In, Rescinds Car Tax Increase," *Los Angeles Times*, November 18, 2003.

3. For a full list of the transition team, see "Transition: 68 to Advise Incoming Administration," *San Jose Mercury News*, October 10, 2003.

4. The votes were 33–0 in the senate and 64–9 in the assembly. Six Latino senators abstained.

5. Said the head of the Mexican American Political Association, "The Latino and immigrant communities are clearly disappointed by the Senate's decision to repeal SB 60." Quoted in "State Moves Toward Repeal of License Law," *Los Angeles Times*, November 25, 2003.

6. "Suspension of Pending Davis Rules Could Affect Many in State," *Los Angeles Times*, November 19, 2003.

7. "Workers' Comp Crisis Worsens," *Los Angeles Times*, May 25, 2003.

8. "Schwarzenegger's Progress on Plans for First 100 Days as Governor," *San Francisco Chronicle*, December 10, 2003.

9. "Schwarzenegger Restores Funds to Cities, Disabled," *San Francisco Chronicle*, December 19, 2003.

10. "Testing Time Begins for Schwarzenegger," *New York Times*, November 17, 2003.

11. Quoted in "Governor Stumping for Budget Plans," *San Jose Mercury News*, December 3, 2003.

12. "Governor Signs Bond, Balanced Amendment Package," *San Francisco Chronicle*, December 12, 2003.

13. "Schwarzenegger Blinks in Budget Confrontation with Democrats," *Sacramento Bee*, December 12, 2003.

14. "The Inaugural Address" (full text), *San Jose Mercury News*, November 18, 2003.

Index

Abernathy, Mark, 51, 97
Abortion rights, 8, 44, 57, 73, 91–92, 98
Adams, Iris, 80
Affirmative action, 6, 57
AFL-CIO, 101
African American constituency, 99, 136
AGI Management Corporation, 103
Agricultural industry, 9–10, 101
Alexander, Kim, 107
American Civil Liberties Union (ACLU),
   66, 138
American Independent Party, 14
American Media, 125
American Sterling Corp., 105
Angelides, Phil, 160
Anti-recall efforts, 58
Approval ratings, 47, 50, 56
Arduin, Donna, 157
Arnold, Tom, 133
Automobile emissions, 8

Bader, Tom, 58
Bajwa, Vik, 80
Ballot box budgeting, 6–7
Bilingual education ban, 6
Blagojevich, Rod, 16
Blanket ("open") primary system, 148
Bono, Sonny, 8
Bowen, Debra, 34
Boxer, Barbara, 76
Bradley, Tom, 39
Brokaw, Tom, 7, 126
Brown, Jerry, 39–40, 82, 147
Brown, Pat, 82
Brown, Willie, 41, 99, 155
Brulte, Jim, 17, 31–32, 132
Budget deficit, 21–24
  as major public issue, 85–88
  Schwarzenegger's action on, 157
Buffett, Warren, 87, 104, 120
Burton, John, 102

Bush, George W., 9, 59–60, 118, 134, 152
Bush, Jeb, 157
Business interests, 103–106
Bustamante, Cruz, 41, 65, 80, 98, 143
  debates of, 118
  election campaign and, 114–115, 134
  Indian gaming interests and, 43, 106–107
  issue positions, 84–94
  Latino groups, 100
  liberal groups and, 98–99
  as replacement candidate, 14, 74–76, 78,
    109, 146
  television advertising, 114
  union groups, 101–102

Cain, Bruce, 62
California
  direct democracy in, 4–7
  diversity in, 11–12, 96
  economy of, 9–11
  future of politics in, 148–149
  personality politics in, 82, 108–109
  political history of, 3–4
  powers of governors, 16
  regional interests of, 12
  structural challenges of governing, 143–148
  trend setting by, 6–8, 14–15, 151–152
California Business Roundtable, 58–59, 103
California Chamber of Commerce, 103,
    105–106, 116, 159
California Common Cause, 107–08
California Federation of Labor, 59
California Labor Federation, 113
California Manufacturers and Technology
    Association, 105
California Medical Association, 44
California Professional Firefighters Union, 58
California State Assembly, 32
California State Association of Electrical
    Workers, 102
California State Elections Code, 14

California State Employees Association, 132
California State University, 7
California Supreme Court, 3, 36, 66
California Teachers Association, 43, 102
California Voter Foundation, 107
Camejo, Peter, 77–78, 80, 85, 87–88, 90–92, 118, 122
Campaign contributions and spending, 44, 56–57, 69, 107–108, 130
Campaign finance reform, 94
Capital punishment, 43
Car tax, 85–86, 94, 159
Celebrity politics, 111, 123–126, 142
Chavez, Cesar, 8
Checchi, Al, 39
Cheney, Dick, 19
Chevron, 103
Chiampou, Ken, 56
Clark, Wesley, 152
Clinton, Bill, 38, 48–49, 73, 92, 109, 113, 123, 133, 152
Clinton, Hillary, 49
Connerly, Ward, 90
Conservative groups, 97–98
  antitax activists, 97
  religious conservatives, 97
Contract with America, 60
Contreras, Miguel, 102
Cook, Mary "Mary Carey," 80
Costa, Ted, 51, 54–57, 62, 69, 97, 146
Coupal, Jon, 54, 97
Criminal sentencing, 54

Davis, Gray, 3–4, 13–16, 65
  1998 governor's election, 39–40
  2002 governor's election, 44–46, 50
  approval ratings of, 47, 50, 56
  biography of, 38–41
  budget deficit, 21–24, 31–32, 34, 46–47, 150
  business interests and, 103–104, 106
  Bustamante and, 41
  economic recession and, 20–21
  failed leadership of, 30, 94
  fundraising efforts of, 44
  Indian casinos issue, 42–43
  issue positions, 84, 132
    economy and budget, 85–87
    education, 88
    immigration, 90–91
    other issues, 93–94, 103
    social issues, 91–93
  legislature and, 36–37
  personality factors, 24–25, 38, 40–41, 133–134, 143
  popularity polls, 19
  power crisis, 16–19, 44, 47, 88–89

Davis, Gray (continued)
  recall election
    debates, 118
    efforts against recall, 58–59
    endorsements and supporters, 132–133
    factors in defeat, 130–139, 149–150
    interest groups, 97, 99
    lessons of defeat, 141–143
    media campaign, 113–116
    voting results of, 81, 136–138
  television advertising, 113–114
  union support of, 101
  vision and style of governing, 41–46
Davis, Sharon Ryer, 39, 133
Dean, Howard, 55, 152
Debates, 117–119, 135
Defense contractors, 9–10
Democratic Party, 3, 98
  2002 election, 44–45
  budget stalemate, 31–32
  CA's ultraliberalism, 13
  get-out-the-vote (GOTV) drives, 112
  labor groups, 100–103
  redistricting, 35–36
Deukmejian, George, 40, 42, 82
Direct democracy, 4–7, 14, 82
  initiative process, 6
  tools of, 5
Direct mail campaigns, 113, 115
Direct voter involvement, 7
Diversity, 11–12, 96
  immigration as voter issue, 89–91, 132
Doak, David, 130
Domestic partner rights, 9, 98
Door-to-door canvassing, 112
Dot-com bubble, 10, 47

E-mail lists, 55
Eastin, Delaine, 41
Eastwood, Clint, 8
Economic recession, 20–21
  as major public issue, 85–88
  personal income growth, 20
Education, 7, 45–46
  spending on, 22
  as voter issue, 88, 144
Ehrlich, Robert, Jr., 16
Election campaigns, 112–116
  celebrity politics, 123–126, 142
  debates, 117–119
  direct mail, 113
  focus groups, 116
  mass media and, 112–113
  news media and, 119–123
  public opinion polls, 116–117

Election campaigns (*continued*)
"retail" politics, 112
talk radio and, 55–56, 86, 111, 113, 123
television advertising, 113–115
"wholesale" techniques, 112–113
Election cycle, 65
Electricity deregulation (1996), 17
Energy crisis; *see* Power crisis
English as official state language, 54
Enron, 89, 103
Entertainment industry/media, 9–10, 111,
123–124, 142
Environmental protection, 8, 19, 46, 93–94,
98, 100

Family planning funding, 8
Farm workers movement, 8
Federal Energy Regulatory Commission
(FERC), 19
Feinstein, Dianne, 39, 58, 79, 109,
112–114
Feliz, John, 117
Flynt, Larry, 80
Focus groups, 116
Frazier, Lynn, 7

Gallagher, Leo, 80
Gann, Paul, 51, 53–54
Gay rights, 8–9, 91–92, 98
Get-out-the-vote (GOTV) drives, 112
Gilliard, Dave, 57, 155
Gingrich, Newt, 60
Giuliani, Rudy, 114
Globalization, 10
Glover, Danny, 99, 108
Goldberg, Lenny, 54
Gore, Al, 9, 109, 118
Gosse, Rich, 80
Grammer, Kelsey, 133
Grange Party, 3
Green Party, 14, 77
Gruener, Garrett, 80, 108
Gun control, 91–92, 97–98

Harman, Jane, 39
Harris, Todd, 120
Health care issues, 93, 102–103, 106
Healthy Families program, 22
Hedgecock, Roger, 55, 86
Hertzberg, Bob, 155
Hickey, John J. "Jack," 80
High-tech industry, 9, 20, 23, 47
Hoffman, Dustin, 133
Hogue, Eric, 55
Hollywood community, 133
Homeland security, 23
Hoover Institution, 104

Huffington, Arianna, 79–80, 85, 87–88, 90,
92, 94, 99, 108, 118, 133
Huffington, Michael, 13, 79

Illegal immigrants, 6, 41
Immigration, as voter issue, 89–91, 132
Impeachment process (federal), 65
Indian casinos, 42–43, 76, 93, 105–06
Initiative process, 5, 7, 96
Initiative and Referendum Institute, 151
Institute of Governmental Studies (UC
Berkeley), 62
Interest groups, 6, 12–3, 30, 73, 96–97
business interests, 103–106
campaign spending, 108
conservative groups, 97–98
donor groups, 107–108
Indian tribes, 106–107
labor groups, 100–103
liberal groups, 98–100
International press, 125–126, 155
Internet, 111, 123
as news site, 123
recall websites, 50, 55–56, 167–169
Issa, Darrell, 13, 56–59, 62, 69–70, 86, 104,
108, 115, 151

Jackson, Jesse, 99, 133
Jarvis, Howard, 51, 53
Jennings, Peter, 126
Johnson, Hiram, 5, 82, 93
Justice Department, 66

Kaloogian, Howard, 51, 54
Kelly, Walt, 149
King, Larry, 108, 123
Knox, Jim, 107
Kobylt, John, 56

Labor Federation, 102
Labor unions, 100–103, 110, 112, 136
endorsements and, 132–133
labor mailers, 113
Latino vote, 75–76, 90, 99–100, 110, 115,
132, 136
Lay, Ken, 89
League of Conservation Voters, 98
League of Women Voters, 117
Legislature, 29–37
absolute two-thirds vote problem, 30–32,
144
gridlock of, 36, 48, 60, 144, 155
Latino Caucus, 45
limitations of, 29–30
lobbyists and, 34
partisan struggles of, 30
redistricting and, 30, 35, 51, 145–146

Legislature (*continued*)
  safe districts and conflicts, 35–36
  state budget stalemate, 30–31
  term-limits paradox, 32–35, 145
Lehane, Chris, 58
Leno, Jay, 4, 71–73, 123–124, 142
Letterman, David, 73, 124, 141
Liberal groups, 98–100
  African American constituency, 99
Libertarian Party, 14, 80
Limbaugh, Rush, 73
Lincoln Club of Orange County, 59, 79, 97
Lobbying/lobbyists, 34, 105
Lockyer, Bill, 10, 135, 146
Los Angeles County Labor Federation, 102
Lowe, Rob, 108, 133, 154
Lungren, Dan, 40
Luntz, Frank, 60

McClintock, Tom, 14, 98, 106, 109, 122,
    131, 160
  as conservative candidate, 76–79,
    146–147
  election debates, 118–119
  issue positions of, 85–88, 90, 92–94
  media campaign of, 115–116, 123
McNally, Ray, 115
Maldonado, Abel, 109
Manufacturing industry, 9–10, 20
Marijuana, medical use of, 6
Marinucci, Carla, 72
Martorana, Gino, 81
Mass media campaigns, 112–113
Media, 111–112; *see also* Election
    campaigns
  celebrity politics, 123–126, 142
  entertainment media, 111, 123–124
  international press, 125–126, 155
  news media (traditional), 111, 119, 121
  talk radio, 55–56, 86, 111, 123
  traditional campaign media, 111
Mexican American Political Association
    (MAPA), 100
Michael, Jay, 34
Military manufacturing, 9–10
Miller, Dennis, 133
Morgan, Melanie, 50, 55–56
Movimiento Estudiantil Chicano de Aztlan
    (MEChA), 100
Mulholland, Bob, 54
Murkowski, Frank, 16
Murphy, George, 8

National Association for the Advancement
    of Colored People (NAACP), 99, 138
National Organization for Women (NOW), 99
National Rifle Association (NRA), 97

Natural Law Party, 14, 80
New Majority, 104
News media, 111, 119, 121
Nixon, Richard, 8

O'Brien, Conan, 124
O'Connell, Jack, 109
O'Connor, Barbara, 119
Official ballot, 68
Oil industry, 9
Oprah Winfrey Show, 73, 123, 135

Pacific Gas and Electricity (PG&E), 17, 103
Palast, Greg, 89
Panetta, Leon, 152
Partisan politics, 32, 152
"Pay-to-play," 22, 43, 73, 93, 96, 106, 150
Pelosi, Nancy, 58
People's Advocate, 51, 53–54, 56, 97
Personal income growth, 20
Personality politics, 82, 108–109
Petition process, 5
Political consultants, 6
Political parties, 12–14
  ideological components of, 13
  internal tensions of, 13
  smaller parties, 14
  state party organizations, 13
  weak organizational structure of, 146–148
Political reform, 93
Politics, celebrity politics, 111, 123–126, 142
Power crisis, 16–19, 44, 47, 103
  deregulation (1996), 17
  as voter issue, 88–89
Prady, Bill, 81
Primary elections, 13, 148
  blanket ("open") system, 148
Prisons, 22, 46
Professional Engineers of California, 102
Progressives, 3, 5, 14
Property tax reform; *see* Proposition 13
Proponent's Grounds for Recall and the
    Governor's Response, 52
Proposition 4 (government spending
    increases), 54
Proposition 13 (property tax reform), 3, 6,
    14, 51, 53–54, 59, 80, 87, 120, 144
Proposition 34 (political contribution
    limits), 76, 107
Proposition 49 (after-school programs), 72
Proposition 54 (racial privacy), 90–91,
    100–102, 107, 114, 147
Proposition 140 (term limits), 32
Proposition 187 (benefits for immigrants),
    6, 41, 90–91, 100
Proposition 209 (affirmative action), 6, 57,
    100

Public opinion polls, 116–117, 122, 142
  exit polls, 135–136, 138
  external polls, 116
  tracking polls, 116
Public Policy Institute of California (PPIC), 83, 116–117
Public Utilities Commission (PUC), 18
Pulaski, Art, 59, 101

Race riots, 8
Radio media; *see* Talk radio
Ragone, Peter, 58
Ranken, Christopher, 81
Rather, Dan, 126
Reagan, Ronald, 8, 49, 53, 82, 147
Recall election, 5, 7, 49
  business interests, 103–106
  conservative groups, 97–98
  court action on, 66–67
  election order, 60–62
  fairness issue, 130
  Indians groups, 106–107
  instigators and visionaries, 50–54
  Internet websites for, 50, 55–56, 167–168
  issue concerns of voters, 83–95
    candidate position on, 83–85
    car tax, 85–86, 94, 159
    economy and budget, 85–88
    education, 88, 144
    energy crisis, 88–89
    immigration, 89–91
    leadership issue, 94–95
    other issues, 93–94
    social issues, 91–93
  labor groups, 100–103
  liberal groups, 98–100
  major candidates, 71–80; *see also specific candidates*
    Bustamante, Cruz, 74–76
    Camejo, Peter, 77–78
    Huffington, Arianna, 79–80
    McClintock, Tom, 76–77
    Schwarzenegger, Arnold, 71–74
    Simon, Bill, 78–79
    Ueberroth, Peter, 79
  as mandate for change, 131
  other candidates, 80–81
  People's Advocate and, 54
  petition circulators, 4, 50–51, 56
  political stability and, 140–153
  proponent's grounds/governor's response for, 52
  qualifying for the ballot, 49–50
  recall and replacement election, 67
  requirements for replacement ballot, 65
  Rescue California website, 57–58, 153
  results of, 81

Recall election (*continued*)
  as right-wing efforts, 49, 56, 59
  signatures by county, 61
  talk radio, 55–56, 86, 111, 113, 123
  timeline of, 65, 163–166
  as trendsetter scenario, 151–152
  two-part ballot, 67
  as two-person race, 134–135
  use of, 6–7
  voting groups, 135–138
Recall Gray Davis Committee, 51
Recall timeline, 163–166
Recall websites, 50, 55–57, 167–168
Redistricting, 30, 35, 51, 145–146
Referendum, 5
Religious conservatives, 97
Renz, Reva Renee, 81
Republican Party, 8
  2002 election, 44–45
  budget stalemate, 31–32
  CA's ultraconservatism, 13, 45
  conservative-moderate schism, 146–147
  recall election and, 54, 59–60, 70
  redistricting, 35–36
Rescue California website, 57–58, 153
Ridley-Thomas, Mark, 153
Right to privacy, 8
Riordan, Richard, 44–45, 71, 104, 135, 148
Rove, Karl, 59
Russo, Sal, 50, 54–55, 62

Salazar, Roger, 69
Salladay, Robert, 134
San Diego Gas & Electric, 18
Sarkozy, Nicolas, 125
Savage, Michael, 55
Schell, Orville, 124
Schnur, Dan, 120
Schwarzenegger, Arnold, 4, 8, 13–14, 59–60, 79–80
  bus tour, 121
  business interests and, 104–106
  campaign strategy of, 72–74, 95, 119, 122–123
  celebrity politics of, 124–126, 142
  conservative groups and, 97–98
  debate participation, 118–119
  election victory of, 131, 137–138
  feminist groups and, 99
  focus groups, 116
  as governor, 148–149, 154–155
    activism and campaign promises, 155
    bully pulpit political style, 158–160
    early compromises, 157–158
    early victories/defeats, 156–157
  issue positions, 84
    economy and budget, 85–87

Schwarzenegger, Arnold
  issue positions *(continued)*
    education, 88
    energy crisis, 89
    immigration, 90–91
    other issues, 93–94
    social issues, 92
  major endorsements of, 122
  media campaign and coverage, 113,
    115, 120–121
  mystique of celebrity, 71–74, 82,
    108–109
  sexual harassment allegations, 70–71,
    73, 92, 135
  structural issues as governor, 143–144
  television advertising, 113–114
  top donor groups for, 105
Senate Bill (SB) 2 (health care coverage),
  93, 102–103, 106
Separation of powers concept, 42
September 11th attacks, 10, 23
Service industries, 9–10
Shelley, Kevin, 51, 60, 65–66, 157
Shelton, George, 150
Shriver, Maria, 72–73, 92, 99, 109,
  122–123, 154
Shultz, George, 104, 120
Sierra Club, 98
Silicon Valley, 9–10
Simon, Bill, 14, 19, 44–45, 50, 71,
  77–79, 104, 131
Sipple, Don, 119
Skelton, George, 143
Social issues, as voter concern, 91–93
Sorkin, Aaron, 133
South, Davis, 130
South, Gary, 58, 122
Southern California Edison, 18, 103
Southern Pacific Railroad, 5, 30
Spanos, A. G., 105
Special interests; *see* Interest groups
Starr, Kevin, 148
Steel, Shawn, 56
Stern, Howard, 123, 142
Stewart, Jon, 124
Stoos, John, 51
Streisand, Barbra, 113, 133
Student movement, 8
Stutzman, Rob, 109
Sundheim, Duf, 59

Talk radio, 55–56, 86, 111, 113, 123
Taxpayer revolt, 6–7, 51
Taxpayers Against the Governor's Recall, 58
Television advertising, 113–115
Term-limits, 3, 6, 32–35, 51, 54, 96, 145
Terry, Dan, 58
Third party candidate, 77
"Three strikes" initiative, 6, 43
Tonight Show, 4, 72, 123–124
Torres, Art, 133
Tourism, 9–10
Town meetings, 112
Two-part ballot, 67

Ueberroth, Peter, 14, 79, 108, 118
Unemployment, 20–21
Union activism, 101
Unions; *see* Labor unions
United Farm Workers, 100
U.S. House of Representatives, 35, 57
U.S. Supreme Court, 35
University of California, 7

Van de Kamp, John, 42
Vandeventer, James M., Jr., 81
Vasconcellos, John, 34
Vehicle license fee, 85–86, 94, 159
Victim's Bill of Rights, 54
Villaraigosa, Antonio, 22
Voting groups, 135–138
Voting Rights Act, 66

Walsh, Sean, 109, 142
Walters, Dan, 34, 43, 46, 158
Warren, Earl, 82, 147
Waters, M. Dane, 151
Watts riots, 8
Website for recall, 50, 55–57, 167–168
Weider, Joe, 125
Weintraub, Daniel, 104
Westly, Steve, 77
White, Bob, 109
Williams, Mark, 55
Wilson, Pete, 8, 17, 39–40, 82, 85, 89–90,
  104, 109
Winfrey, Oprah, 108, 123, 142
Women's groups, 98, 110
Workingmen's Party, 3
Workmen's compensation, 87

Zucchino, David, 71

About the Authors

Larry N. Gerston is professor of political science at San Jose State University and the author or co-author of seven books on state and national politics. His *Public Policy Making in a Democratic Society: A Guide to Civic Engagement* has, been heralded as a breakthrough primer for citizen involvement. Likewise, his *California Politics and Government: A Practical Guide* (co-authored with Terry Christensen) has sold more than 100,000 copies as the most popular book in its field. A native Californian, Gerston has worked for a county supervisor and a state legislator. For the past twenty-three years, he has been the political analyst at NBC11 in the San Francisco Bay area. In addition, Gerston has published *The Costco Experience: An Unofficial Survivor's Guide*, a hands-on, tongue-in-cheek book about the world of warehouse shopping.

Terry Christensen is professor of political science at San Jose State University and the author or co-author of six books and many newspaper op-ed pieces. He is frequently consulted by local and national media concerning political issues in California and Silicon Valley. He is currently at work on new editions of *Reel Politics: American Political Movies from* Birth of a Nation *to* Platoon, and *Local Politics: Governing at the Grassroots* (both forthcoming in 2005). He has served on numerous civic committees and commissions and currently is a member of the Advisory Board of City Year/ Silicon Valley and Program Chair for the Commonwealth Club/Silicon Valley. In 1998, he was named San Jose State University's Outstanding Professor.